Lecture Notes in Mathematics

Edited by A. Dold and B. Eckmann

538

Gerd Fischer

Complex Analytic Geometry

Springer-Verlag
Berlin · Heidelberg · New York 1976

Author
Gerd Fischer
Mathematisches Institut
der Universität München
Theresienstraße 39
8000 München 2/BRD

Library of Congress Cataloging in Publication Data

Fischer, Gerd, 1939-
Complex analytic geometry.

(Lecture notes in mathematics ; 538)
1. Functions of several complex variables.
2. Analytic spaces. 3. Fiber spaces (Mathematics)
I. Title. II. Series: Lecture notes in mathematics (Berlin) ; 538.
QA3.L28 vol. 538 [QA331] 510'.8s [515'.94]
76-27692

AMS Subject Classifications (1970): 32BXX, 32CXX, 32J10, 32LXX

ISBN 3-540-07857-6 Springer-Verlag Berlin · Heidelberg · New York
ISBN 0-387-07857-6 Springer-Verlag New York · Heidelberg · Berlin

Printed in Germany
Printing and binding: Beltz Offsetdruck, Hemsbach/Bergstr.

PREFACE

These Lecture Notes arose from courses held at the Universities of Regensburg, Frankfurt and Munich. The aim of the courses was to present a survey of the fundamental concepts and results of "complex analytic geometry" (i.e. the theory of functions of several complex variables) and to approach the current state of research in this field.

The choice of the material was governed by the idea that the presentation should center on (not necessarily reduced) complex spaces and their holomorphic maps. Progressing from the fundamentals of this theory to problems of current interest is a particularly long and arduous task. Bearing in mind the reader's patience, the length of the notes and last but not least the energy of the author, several basic but voluminous methods and proofs have only been cited or sketched, particularly if they are easily accessible in the literature (e.g. [1], [3], [11], [40], [44], [50], [51], [55], [56], [57], [72], [107], [111], [150]). Among these topics are the theory of sheaves, the theory of local analytic algebras and their dimensions, the theory of holomorphic functions in $\mathbb{C}^n$ including the elementary theory of analytic sets, and finally the fundamental coherence theorems of OKA, CARTAN, GRAUERT as well as Theorems A and B of CARTAN-SERRE.

Although these notes are far from being self-contained, the intention was to provide at least the beginnings of a systematic presentation within the framework of each chapter. Hopefully they may serve as a partial substitute for a compendium of the "elements of analytic geometry" which is long overdue.

My sincere thanks go to all who attended the lectures for their suggested improvements and their endurance, to many colleagues for their helpful comments, to Inge Pfeilschifter and Ghislaine Maurer for preparing the camera-ready manuscript, to Camilla Aman for proof-reading and to Joseph Maurer for sketching the figures.

CONTENTS

Chapter 3. DEGENERACIES AND FLATNESS

Chapter 4. MODIFICATIONS AND MEROMORPHIC FUNCTIONS

•••

Chapter 0. Basic Notions

0.1. We assume that all rings occuring here are commutative and have a unit element. Ring homomorphisms are always supposed to respect the unit elements. For the basic notions of sheaf theory we refer to [40].

A *ringed space* is a pair $X = (|X|, \mathcal{O}_X)$ consisting of a topological space $|X|$ and a sheaf of rings $\mathcal{O}_X$ on $|X|$.

It is called a *local ringed space*, if for every $p \in |X|$ the *stalk* $\mathcal{O}_{X,p}$ is a local ring. Its maximal ideal is denoted by $m_{X,p}$.

A local ringed space X is called a $\mathbb{C}$-ringed space, if furthermore $\mathcal{O}_X$ is a sheaf of $\mathbb{C}$-algebras and for every $p \in |X|$ there is an isomorphism

$$\mathcal{O}_{X,p}/m_{X,p} \to \mathbb{C}$$

of $\mathbb{C}$-algebras.

For simplicity we shall mostly write X instead of $|X|$, i.e. $X = (X, \mathcal{O}_X)$ by abuse of notation.

If $U \subset X$ is an open subset, then U together with the restriction

$$\mathcal{O}_U := \mathcal{O}_X|U$$

is again a ringed space.

If X is a ringed space, any sheaf F of $\mathcal{O}_X$-modules is called an $\mathcal{O}_X$-*module* or a *sheaf of modules over* X. For $U \subset X$ open, $F(U)$ denotes the $\mathcal{O}_X(U)$-module of *sections* of F over U. If $f \in F(U)$, its *germ* at $p \in U$ is denoted by f_p. If $f \in \mathcal{O}_X(U)$ and $\mathcal{O}_{X,p}$ is local,

$$f(p) := f_p + m_{X,p} \in \mathcal{O}_{X,p}/m_{X,p}$$

is called the *value* of f in p.

0.2. Let ringed spaces X and Y and a continuous map $\varphi: |X| \to |Y|$ be given. If F is an $\mathcal{O}_X$-module, the presheaf

$$V \mapsto F(\varphi^{-1}(V)), \quad V \subset Y \text{ open},$$

on Y is a sheaf. We denote it by $\varphi_* F$ and call it the *direct image* of F. It is a $\varphi_*\mathcal{O}_X$-module.

If $\alpha: F \to G$ is a homomorphism of $\mathcal{O}_X$-modules, we define

$$\varphi_*\alpha : \varphi_* F \to \varphi_* G$$

by $(\varphi_*\alpha)(V) := \alpha(\varphi^{-1}(V))$. $\varphi_*\alpha$ is a homomorphism of $\varphi_*\mathcal{O}_X$-modules.

If G is an $\mathcal{O}_Y$-module, we denote by $\varphi^{-1}G$ the *topological inverse image* of G, which is a sheaf on X uniquely determined by the property

$$\varphi^{-1}G_p = G_{\varphi(p)}$$

for all $p \in X$. $\varphi^{-1}G$ is a $\varphi^{-1}\mathcal{O}_Y$-module. If we consider a sheaf as an "espace étalé" [40] we can define $\varphi^{-1}G$ as the fibre product

$$X \times_Y G = \{(x,g) \in X \times G\colon g \in G_{\varphi(x)}\}.$$

<u>0.3.</u> A *morphism* $\varphi\colon X \to Y$ *of ringed spaces* $X = (|X|, \mathcal{O}_X)$ and $Y = (|Y|, \mathcal{O}_Y)$ is a pair $\varphi = (|\varphi|, \tilde{\varphi})$ consisting of a continuous map

$$|\varphi|\colon |X| \to |Y|$$

and a homomorphism

$$\tilde{\varphi}\colon \mathcal{O}_Y \to |\varphi|_*\mathcal{O}_X$$

of sheaves of rings on Y. Instead of $|\varphi|_*\mathcal{O}_X$ we shall write $\varphi_*\mathcal{O}_X$ for simplicity.

The ring homomorphism

$$\tilde{\varphi}_p\colon \mathcal{O}_{Y,\varphi(p)} \to \mathcal{O}_{X,p}$$

for $p \in X$ is defined as the composition of the canonical homomorphisms

$$\mathcal{O}_{Y,\varphi(p)} \to (\varphi_*\mathcal{O}_X)_{\varphi(p)} \to \mathcal{O}_{X,p}.$$

In case X and Y are *local* ringed spaces a morphism by definition has to satisfy the condition that

$$\tilde{\varphi}_p\colon \mathcal{O}_{Y,\varphi(p)} \to \mathcal{O}_{X,p}$$

be *local* for every $p \in X$, i.e.

$$\tilde{\varphi}_p(m_{Y,\varphi(p)}) \subset m_{X,p}.$$

A *morphism of* $\mathbb{C}$*-ringed spaces* X and Y is a morphism of ringed spaces where $\tilde{\varphi}$ is furthermore a homomorphism of sheaves of $\mathbb{C}$-algebras. In this case $\tilde{\varphi}_p$ is automatically local (see [52] 9 - 11).

The composition of morphisms is constructed in an obvious manner: If $\varphi\colon X \to Y$ and $\psi\colon Y \to Z$ are two morphisms, we define $|\psi\circ\varphi| = |\psi|\circ|\varphi|$. Applying $|\psi|_*$ to $\tilde{\varphi}$ yields

$$|\psi|_*\tilde{\varphi}\colon |\psi|_*\mathcal{O}_Y \to |\psi|_*(|\varphi|_*\mathcal{O}_X) = (|\psi|\circ|\varphi|)_*\mathcal{O}_X$$

an we define $(\widetilde{\psi\circ\varphi}) = (|\psi|_*\tilde{\varphi})\circ\tilde{\psi}$.

We obtain the categories of ringed spaces, local ringed spaces and $\mathbb{C}$-ringed spaces. Now *monomorphisms* and *epimorphisms* are defined by the usual cancellation properties, and a morphism $\varphi: X \to Y$ is called an *isomorphism*, if there exists a morphism $\psi: Y \to X$ such that $\psi \circ \varphi = id_X$ and $\varphi \circ \psi = id_Y$.

<u>0.4.</u> <u>Lemma.</u> Let $\varphi: X \to Y$ be a morphism of $\mathbb{C}$-ringed spaces.

i) If $|\varphi|$ is injective and $\widetilde{\varphi}_p$ is surjective for every $p \in X$, then φ is a monomorphism.

ii) If $|\varphi|$ is surjective and $\widetilde{\varphi}_p$ is injective for every $p \in X$, then φ is an epimorphism.

iii) φ is an isomorphism if and only if $|\varphi|$ is a homeomorphism and $\widetilde{\varphi}_p$ is an isomorphism for every $p \in X$.

The proof is straightforward.

<u>0.5.</u> Let $X = (|X|, \mathcal{O}_X)$ be a ringed space and F sheaf of modules over X. F is called of *finite type* (resp. *locally free*) if for every point p of X there is an open neighbourhood U and an epimorphism (resp. isomorphism)

$$\alpha: \mathcal{O}_U^k \to F|U.$$

In case α is an epimorphism we define

$$s_1 := (1,0,\dots,0), \dots, s_k := (0,\dots,0,1) \in \mathcal{O}_U^k(U) \text{ and}$$

$$f_1 := \alpha(U)(s_1), \dots, f_k := \alpha(U)(s_k) \in F(U);$$

then $f_{1,x}, \dots, f_{k,x} \in F_x$ generate F_x over $\mathcal{O}_{X,x}$ for every $x \in U$.

$$R := Ker(\mathcal{O}_U^k \to F|U) \subset \mathcal{O}_U^k$$

is called the *sheaf of relations* of $f_1, \dots, f_k$.

F is called of *finite presentation*, if for every point p of X there is an open neighbourhood U and an exact sequence

$$\mathcal{O}_U^l \to \mathcal{O}_U^k \to F|U \to 0.$$

This means that F is of finite type and in addition the sheaves of relations defined as above are of finite type.

F is called *coherent* (or more precisely $\mathcal{O}_X$-coherent), if

i) F is of finite type and

ii) for every open $U \subset X$ and every homomorphism

$$\mathcal{O}_U^k \to F|U$$

the kernel is of finite type over U.

ii) is equivalent to

ii') for every open $U \subset X$ and any $f_1,\dots,f_k \in F(U)$ the sheaf of relations $R \subset \mathcal{O}_U^k$ is of finite type over U.

Example. If $X = (\mathbb{C},\mathcal{O}_{\mathbb{C}})$, where $\mathcal{O}_{\mathbb{C}}(U)$ denotes the ring of holomorphic functions on $U \subset \mathbb{C}$, $A := \{1,\frac{1}{2},\frac{1}{3},\dots\}$ and F is defined by

$$F(U) = \{f \in \mathcal{O}_{\mathbb{C}}(U): f|(U \cap A) = 0\},$$

then F is not of finite type.

Remark. If X is a ringed space and F is an $\mathcal{O}_X$-module of finite type, then the *support*

$$\operatorname{supp} F := \{x \in X: F_x \neq 0\} \subset X$$

is a closed subset.

Proof. If $U \subset X$ is open and $f \in F(U)$, then obviously $\{x \in U: f_x = 0\} \subset U$ is open. For $p \in X$ take an open neighbourhood U and $f_1,\dots,f_k \in F(U)$ whose germs at x generate F_x for every $x \in U$. Then

$$(\operatorname{supp} F) \cap U = \bigcup_{i=1}^{k} \{x \in U: f_{i,x} \neq 0\},$$

and the assertion follows.

0.6. Lemma. Let X be a ringed space.

a) Let F be a coherent $\mathcal{O}_X$-module and $G \subset F$ a submodule. Then G is coherent if and only if it is of finite type.

b) Let

$$0 \to F' \to F \to F'' \to 0$$

be an exact sequence of three $\mathcal{O}_X$-modules. If two of the modules are coherent, then the third one is also coherent.

c) If $\alpha: F \to G$ is a homomorphism of coherent $\mathcal{O}_X$-modules, then $\mathcal{K}er\ \alpha$ and $\mathcal{C}oker\ \alpha$ are coherent $\mathcal{O}_X$-modules.

d) If F and G are coherent $\mathcal{O}_X$-modules, then $F \times G$ and $\mathcal{H}om_{\mathcal{O}_X}(F,G)$ are coherent $\mathcal{O}_X$-modules.

For the proof we refer to [128] (or e.g [44]).

Obviously every coherent module is of finite presentation. As an immediate consequence of the lemma we obtain:

Corollary 1. Let X be a ringed space such that $\mathcal{O}_X$ is coherent. Then any $\mathcal{O}_X$-module of finite presentation is coherent.

Part c) of the lemma and the remark of 0.5 imply:

Corollary 2. Let F,G be coherent O_X-modules and assume a homomorphism $\alpha: F \to G$ is given. If $\alpha_p: F_p \to G_p$ is a monomorphism (epimorphism, isomorphism) for $p \in X$ then α has the same property on some open neighbourhood U of p.

Corollary 3. Let $F,F' \subset G$ be coherent submodules. If $F_p \subset F'_p$ for $p \in X$, then $F|U \subset F'|U$ for some open neighbourhood U of p.

The proof is left to the reader.

0.7. Lemma. Let X be a ringed space and assume O_X is O_X-coherent. If $I \subset O_X$ is a coherent ideal and F is an O_X/I-module we consider F as an O_X-module via $O_X \to O_X/I$. Then F is O_X/I coherent if and only if F is O_X-coherent.

In particular O_X/I is O_X/I-coherent.

Proof. We make use of Lemma 0.6. Since we have the canonical exact sequence

$$0 \to I \to O_X \to O_X/I \to 0,$$

O_X/I is O_X-coherent. It is obvious that F is of finite type with respect to O_X if and only if it is of finite type with respect to O_X/I. Assume F is O_X-coherent and let an exact sequence

$$0 \to K \to ((O_X/I)|U)^n \to F|U$$

over $U \subset X$ open be given. Then K is O_X-coherent as the kernel of a homomorphism between O_X-coherent sheaves. In particular, K is of finite type with respect to O_X, hence of finite type with respect to O_X/I. Assume F is O_X/I-coherent and let an exact sequence

$$0 \to K \to (O_X|U)^n \to F|U$$

be given. By passing from $(O_X|U)^n$ to $((O_X/I)|U)^n$ we see that K is O_X/I coherent. In particular K is of finite type with respect to O_X/I, hence to O_X.

0.8. Proposition. Let O_X-modules F and G and $p \in X$ be given. Let

$$\rho: (Hom(F,G))_p \to \mathrm{Hom}(F_p,G_p)$$

be the canonical homomorphism.

a) If F is of finite type then ρ is injective

b) If F is of finite presentation then ρ is an isomorphism.

For a proof we refer to [128].

0.9. We mention two useful consequences.

Corollary 1. Let F, G be coherent $\mathcal{O}_X$-modules, $p \in X$ and a monomorphism (resp. epimorphism) $\beta: F_p \to G_p$ be given. Then there is an open neighbourhood U of p and a monomorphism (resp. epimorphism)

$$\alpha: F|U \to G|U$$

such that $\alpha_p = \beta$.

This follows from 0.8 together with 0.6,c) and the Remark of 0.5.

Corollary 2. Let X be a ringed space such that $\mathcal{O}_X$ is coherent, $p \in X$.

a) If M is an $\mathcal{O}_{X,p}$-module of finite presentation then there is an open neighbourhood U of p and a coherent $\mathcal{O}_U$-module G such that G_p is isomorphic to M.

b) If F is a coherent $\mathcal{O}_X$-module and $M \subset F_p$ a finitely generated submodule then there is an open neighbourhood U of p and a coherent $\mathcal{O}_U$-submodule $G \subset F|U$ such that $G_p = M$.

This follows from 0.8 and Corollary 1.

0.10. If $\varphi: X \to Y$ is a morphism of ringed spaces and F is an $\mathcal{O}_X$-module then the $\varphi_*\mathcal{O}_X$-module $\varphi_* F$ is also an $\mathcal{O}_Y$-module via

$$\tilde{\varphi}: \mathcal{O}_Y \to \varphi_* \mathcal{O}_X.$$

This homomorphism induces

$$\tilde{\varphi}^{\#}: \varphi^{-1}\mathcal{O}_Y \to \mathcal{O}_X \text{ with } (\tilde{\varphi}^{\#})_p = \tilde{\varphi}_p$$

for $p \in X$ and hence $\mathcal{O}_X$ may be considered as a $\varphi^{-1}\mathcal{O}_Y$ module. If G is an $\mathcal{O}_Y$-module, we may define the *analytic inverse image*

$$\varphi^* G = \varphi^{-1} G \otimes_{\varphi^{-1}\mathcal{O}_Y} \mathcal{O}_X$$

which is an $\mathcal{O}_X$-module.

In this way we get covariant functors φ_* and φ^* from $\mathcal{O}_X$-modules to $\mathcal{O}_Y$-modules and conversely. φ^* is the left adjoint of φ_*, i.e. for any $\mathcal{O}_X$-module F and any $\mathcal{O}_Y$-module G there is a canonical isomorphism

$$\mathrm{Hom}_X(\varphi^* G, F) \to \mathrm{Hom}_Y(G, \varphi_* F).$$

By substituting $F = \varphi^* G$ and $G = \varphi_* F$ we obtain canonical homomorphisms

$$G \to \varphi_*(\varphi^* G)$$

and

$$\varphi^*(\varphi_* F) \to F.$$

For details of this construction we refer to [49] or [54].

We recall some elementary properties of φ_* and φ^*.

a) φ_* is left exact

b) φ^* is right exact

c) $\varphi^* \mathcal{O}_Y^k = \mathcal{O}_X^k$

d) If G is an $\mathcal{O}_Y$-module of finite presentation then $\varphi^* G$ is of finite presentation.

However, in general φ_* is not right exact, φ^* is not left exact and if F is of finite presentation then $\varphi_* F$ need not be of finite presentation.

Examples. 1) φ_* is not right exact.

Take $X = \mathbb{C}^2 \setminus \{o\}$ and $Y = \mathbb{C}$, where $\mathcal{O}_X$ and $\mathcal{O}_Y$ are the ordinary sheaves of holomorphic functions. Denote by

$$\varphi: \mathbb{C}^2 \setminus \{o\} \to \mathbb{C}, \ (x_1, x_2) \mapsto x_1,$$

the projection and by $X_o \subset X$ the inverse image of $o \in \mathbb{C}$. Then X_o is equal to $\mathbb{C}^*$. If I is the sheaf of ideals of X_o (for details see 0.13), we obtain the canonical exact sequence

$$\mathcal{O}_X \xrightarrow{\alpha} \mathcal{O}_X / I \to 0,$$

but the induced homomorphism

$$\varphi_* \alpha : \varphi_* \mathcal{O}_X \to \varphi_*(\mathcal{O}_X / I)$$

is not surjective. For instance $\frac{1}{x_2}$ is holomorphic on $X_o = \mathbb{C}^*$ but not in the image of $\varphi_* \alpha$.

2) φ^* is not left exact.

Let $X = (\{\cdot\}, \mathbb{C})$ be a point and take $Y = \mathbb{C}$, where $\mathcal{O}_Y$ is the ordinary sheaf of holomorphic functions. Define

$$\varphi: X \to Y \text{ by } \varphi(\cdot) = o.$$

If y is the coordinate function in $\mathbb{C}$, we denote by α the multiplication by y. This yields an exact sequence

$$0 \to \mathcal{O}_Y \xrightarrow{\alpha} \mathcal{O}_Y.$$

But

$$\varphi^* \alpha: \mathbb{C} \to \mathbb{C}$$

is the zero map.

3) $\varphi_* F$ is not of finite type.

Take $X = \mathbb{C}$, Y a point, $F = \mathcal{O}_X$ and

$$\varphi: X \to Y$$

the constant map. Then $\varphi_* \mathcal{O}_X$ is the ring of holomorphic functions in the complex plane, which is not finitely generated over the field $\mathbb{C}$.

<u>0.11.</u> <u>Lemma.</u> Let X be a topological space, $A \subset X$ a closed subset, $\mathcal{O}$ a sheaf of rings on A and F an $\mathcal{O}$-module. Denote by $\iota: A \to X$ the injection. Then F is $\mathcal{O}$-coherent if and only if $\iota_* F$ is $\iota_* \mathcal{O}$-coherent. The proof is very easy.

<u>0.12.</u> By $\mathcal{O}_n$ we denote the sheaf of holomorphic functions in $\mathbb{C}^n$. We state the fundamental

<u>Coherence Theorem of OKA.</u> The sheaf $\mathcal{O}_n$ is $\mathcal{O}_n$-coherent.

For a proof we refer to [107]. By using 0.6 we obtain the

<u>Corollary.</u> a) Every sheaf of ideals $I \subset \mathcal{O}_n$ of finite type is coherent.
b) Every $\mathcal{O}_n$-module F of finite presentation is coherent.

<u>0.13.</u> For the basic properties of analytic sets in open subsets of $\mathbb{C}^n$ we refer to [150].

<u>Lemma.</u> Let $W \subset \mathbb{C}^n$ be an open subset, denote by $\mathcal{O}_W$ the sheaf of holomorphic functions on W and assume F is a coherent $\mathcal{O}_W$-module. Then

$$\operatorname{supp} F = \{x \in W: F_x \neq 0\} \subset W$$

is an analytic subset.

<u>Proof.</u> We first prove that for any coherent sheaf of ideals $I \subset \mathcal{O}_W$ the subset

$$\operatorname{supp}(\mathcal{O}_W/I) \subset W$$

is analytic. For $p \in W$ we take an open neighbourhood U and $f_1, \ldots, f_m \in I(U)$ generating every stalk of I. For $x \in U$ we have

$$\begin{aligned} x \in \operatorname{supp}(\mathcal{O}_W/I) &\Leftrightarrow I_x \neq \mathcal{O}_{W,x} \Leftrightarrow I_x \subset m_{W,x} \\ &\Leftrightarrow f_{1,x}, \ldots, f_{m,x} \in m_{W,x} \\ &\Leftrightarrow f_1(x) = \ldots = f_m(x) = 0, \end{aligned}$$

and the first assertion is proved.

In the general case take $p \in X$ and an open neighbourhood U with an exact sequence

$$\mathcal{O}_U^1 \to \mathcal{O}_U^k \xrightarrow{\alpha} \mathcal{F}|U \to 0.$$

If $\mathcal{R} := \operatorname{Ker} \alpha \subset \mathcal{O}_U^k$, denote by $\mathcal{R}_i \subset \mathcal{O}_U$ for $i = 1,\dots,k$ the i-th component of $\mathcal{R}$. $\mathcal{R}_i$ is a coherent sheaf of ideals and

$$(\operatorname{supp} \mathcal{F}) \cap U = \operatorname{supp}(\mathcal{O}_U/\mathcal{R}_1) \cup \dots \cup \operatorname{supp}(\mathcal{O}_U/\mathcal{R}_k).$$

Since the union of analytic sets is analytic our lemma is proved.

If $W \subset \mathbb{C}^n$ is open and $A \subset W$ is analytic we define a sheaf $\mathcal{J}_A \subset \mathcal{O}_W$ by

$$\mathcal{J}_A(U) = \{f \in \mathcal{O}_W(U) : f|A = 0\}$$

for $U \subset W$ open. $\mathcal{J}_A$ is called the *full sheaf of ideals of* A. We state the deep

<u>Coherence Theorem of CARTAN.</u> If $W \subset \mathbb{C}^n$ is open and $A \subset W$ is analytic, the sheaf $\mathcal{J}_A$ is $\mathcal{O}_W$-coherent [1].

<u>Corollary.</u> If $\mathcal{I} \subset \mathcal{O}_W$ is a coherent sheaf of ideals, then $\operatorname{rad} \mathcal{I} \subset \mathcal{O}_W$ is coherent.

<u>Proof.</u> If $A := \operatorname{supp}(\mathcal{O}_W/\mathcal{I})$ we have

$$\mathcal{J}_A = \operatorname{rad} \mathcal{I}$$

by HILBERTs Nullstellensatz (see for instance [1]).

<u>0.14.</u> Now we can give our basic definitions.

Let $W \subset \mathbb{C}^n$ be open and assume a coherent ideal $\mathcal{I} \subset \mathcal{O}_W$ is given. Then

$$A := \operatorname{supp}(\mathcal{O}_W/\mathcal{I})$$

is an analytic subset of W and $(A, (\mathcal{O}_W/\mathcal{I})|A)$ is a $\mathbb{C}$-ringed space which we call a *local model*.

A *complex space* (or more precisely a $\mathbb{C}$-*analytic space*) is a $\mathbb{C}$-ringed space $X = (|X|, \mathcal{O}_X)$ satisfying the following conditions:

a) $|X|$ is hausdorff.

b) For every $p \in X$ there is an open neighbourhood U of p such that $(U, \mathcal{O}_X|U)$ is isomorphic (as $\mathbb{C}$-ringed space) to some local model.

A complex space X is called *non-singular* at $p \in X$, if there is an open neighbourhood U of p such that $(U, \mathcal{O}_X|U)$ is isomorphic to some local model of the form $(W, \mathcal{O}_W)$, where $W \subset \mathbb{C}^n$ is an open subset.

If X is non-singular at every point, we call it a *complex manifold.*

Example. Denote by z the coordinate function in the complex plane $\mathbb{C}$ and let $I_n \subset O_1$ be the sheaf of ideals generated by z^n for n = 1,2,.... Then

$$\operatorname{supp}(O_1/I_n) = \{o\} \text{ and}$$

$$(O_1/I_n)|\{o\} = \mathbb{C} + \mathbb{C}\varepsilon + \ldots + \mathbb{C}\varepsilon^{n-1},$$

where $\varepsilon^n = 0$. The complex space

$$(\{o\}, \mathbb{C} + \mathbb{C}\varepsilon + \ldots + \mathbb{C}\varepsilon^{n-1})$$

is called an *n-fold point.* For n > 1 it is not a complex manifold. With the notations of 0.13 we have

$$J_{\{o\}} = \operatorname{rad} I_n = I_1$$

for every n.

If X and Y are complex spaces, then any morphism

$$\varphi = (|\varphi|,\tilde{\varphi}): (|X|, O_X) \to (|Y|,O_Y)$$

of $\mathbb{C}$-ringed spaces is called a *holomorphic map.*

A holomorphic map is called *biholomorphic,* if it is an isomorphism of $\mathbb{C}$-ringed spaces (see 0.3).

A complex space Y is called an *open complex subspace* of a complex space X, if

$$|Y| \text{ is an open subset of } |X| \text{ and } O_Y = O_X|Y.$$

In this case we often write $Y \subset X$ by abuse of notation.

Y is called a *closed complex subspace* of X if there is a coherent ideal $I \subset O_X$ such that

$$|Y| = \operatorname{supp}(O_X/I) \text{ and } O_Y = (O_X/I)|Y.$$

In this case there is a canonical holomorphic map determined by the injection, which we denote by

$$Y \hookrightarrow X.$$

Finally a subset A of a complex space X is called *analytic,* if there is a coherent ideal $I \subset O_X$ such that

$$A = \operatorname{supp}(O_X/I).$$

<u>Theorem.</u> The structure sheaf $\mathcal{O}_X$ of every complex space X is coherent.

This is an immediate consequence of the Coherence Theorem of OKA and the lemmas of 0.7 and 0.11.

<u>0.15.</u> If X is a complex space and $U \subset X$ is an open subset, any $f \in \mathcal{O}_X(U)$ is called a *holomorphic function* on U. Such a "function" may be nilpotent, as we have seen in the example of a multiple point (0.14). In order to get such "bad functions" under control we define the *nilradical*

$$\mathcal{N}_X = rad\ \mathcal{O} \subset \mathcal{O}_X$$

as the sheaf of ideals associated to the presheaf

$$U \mapsto \{f \in \mathcal{O}_X(U) : f^k = 0 \text{ for some } k \in \mathbb{N}\}.$$

Then clearly for every $p \in X$ we have

$$\mathcal{N}_{X,p} = \{f \in \mathcal{O}_{X,p} : f^k = 0 \text{ for some } k \in \mathbb{N}\} = \text{rad } \mathcal{O}.$$

By $\mathcal{C}_X$ we denote the sheaf of continuous complexvalued functions on $|X|$. If $U \subset X$ is open, there is a canonical homomorphism

$$\mathcal{O}_X(U) \to \mathcal{C}_X(U),\ f \mapsto \tilde{f},$$

where $\tilde{f}(p) := f(p)$ for $p \in X$ and this defines a canonical homomorphism

$$\mathcal{O}_X \to \mathcal{C}_X.$$

<u>Theorem.</u> Let $(X, \mathcal{O}_X)$ be a complex space. Then
a) the nilradical $\mathcal{N}_X \subset \mathcal{O}_X$ is a coherent ideal and
b) $\mathcal{N}_X = \mathcal{K}er(\mathcal{O}_X \to \mathcal{C}_X)$.

<u>Proof.</u> The question being local we may suppose X is a local model $(A, (\mathcal{O}_W/\mathcal{I})|A)$. With the notations of 0.13

$$\mathcal{N}_X = ((rad\ \mathcal{I})/\mathcal{I})|A = (\mathcal{J}_A/\mathcal{I})|A$$

and hence the Coherence Theorem of CARTAN together with 0.7 and 0.11 implies a).

To prove b) take $p \in X$ and $f_p = F_p + \mathcal{I}_p \in \mathcal{O}_{X,p}$, where $F_p \in \mathcal{O}_{W,p}$. Then

$$\tilde{f}_p = 0 \Leftrightarrow F_p \in \mathcal{J}_{A,p} \Leftrightarrow f_p \in \mathcal{J}_{A,p}/\mathcal{I}_p = \mathcal{N}_{X,p}.$$

We still note a consequence of HILBERTs Nullstellensatz:

<u>Proposition.</u> Let X be a complex space and let $\mathcal{I}, \mathcal{I}' \subset \mathcal{O}_X$ be coherent ideals. If

$\operatorname{supp}(\mathcal{O}_X/I) \supset \operatorname{supp}(\mathcal{O}_X/I')$,

then for any relatively compact open $U \subset X$, there is a $k \in \mathbb{N}$ such that

$I^k|U \subset I'|U$.

In particular for any relatively compact open $U \subset X$ there is a $k \in \mathbb{N}$ such that

$N_X^k|U = 0$.

Proof. Let J and J' be the full sheaves of ideals of the analytic sets $\operatorname{supp}(\mathcal{O}_X/I)$ and $\operatorname{supp}(\mathcal{O}_X/I')$ (see 0.13). By using HILBERTs Nullstellensatz we obtain

$I \subset \operatorname{rad} I = J \subset J' = \operatorname{rad} I'$.

Take $p \in U$. Since $\mathcal{O}_{X,p}$ is noetherian, we can find $r \in \mathbb{N}$ such that $I_p^r \subset I'_p$. Hence

$I^r|V \subset I'|V$

for some open neighbourhood V of p (see 0.6). Since U is relatively compact we may find a covering with a finite number of open sets V as above and the assertion follows.

The second assertion follows from

$\operatorname{supp}(\mathcal{O}_X/N_X) = X = \operatorname{supp}(\mathcal{O}_X/0)$.

0.16. Let $X = (|X|, \mathcal{O}_X)$ be a complex space. Since $N_{X,p} \neq \mathcal{O}_{X,p}$ for every $p \in X$, $\operatorname{supp} \mathcal{O}_X/N_X = |X|$ and

$X_{red} := (|X|, \mathcal{O}_X/N_X)$

is a complex subspace of X, which we call the *reduction* of X.

X is called *reduced*, if $X_{red} = X$.

By using 0.15 we obtain

Lemma. Let X be a complex space. Then the following conditions are equivalent:

i) X is reduced,

ii) $N_X = 0$,

iii) The canonical homomorphism $\mathcal{O}_X \to C_X$ is injective.

For example the reduction of a multiple point is a simple point (see the example in 0.14).

0.17. Lemma. Let $\varphi: X \to Y$ be a holomorphic map and assume closed complex subspaces $X' \hookrightarrow X$ and $Y' \hookrightarrow Y$ are defined by ideals $\mathcal{I} \subset \mathcal{O}_X$ and $\mathcal{J} \subset \mathcal{O}_Y$. Then there exists a uniquely determined holomorphic map $\varphi': X' \to Y'$ such that the diagram

$$\begin{array}{ccc} X & \xrightarrow{\ \varphi\ } & Y \\ {\scriptstyle\iota}\uparrow & & \uparrow{\scriptstyle\kappa} \\ X' & \xrightarrow{\ \varphi'\ } & Y' \end{array}$$

commutes if and only if $\tilde{\varphi}\mathcal{J} \subset \varphi_*\mathcal{I}$.

If we use the notion of the inverse image $\varphi^{-1}(Y')$ of Y' (see 0.27), this condition means that $X' \hookrightarrow \varphi^{-1}(Y')$ is a closed subspace.

φ' is called the *restriction* of φ.

If $X' \subset X$ and $Y' \subset Y$ are open complex subspaces the existence of φ' is equivalent to the condition

$$\varphi(X') \subset Y'$$

for the underlying sets. In the case of closed subspaces this condition is of course necessary but in general by no means sufficient.

Proof. We consider the commutative diagram

$$\begin{array}{ccccc} & & 0 & & 0 \\ & & \downarrow & & \downarrow \\ & & \varphi_*\mathcal{I} & \longleftarrow - & \mathcal{J} \\ & & \downarrow & & \downarrow \\ & & \varphi_*\mathcal{O}_X & \xleftarrow{\ \tilde{\varphi}\ } & \mathcal{O}_Y \\ & & \downarrow & & \downarrow \\ \kappa_*\varphi'_*\mathcal{O}_{X'} = \varphi_*\iota_*\mathcal{O}_{X'} & = & \varphi_*(\mathcal{O}_X/\mathcal{I}) & \xleftarrow{\ \alpha\ } - & \mathcal{O}_Y/\mathcal{J} = \kappa_*\mathcal{O}_{Y'}. \\ & & & & \downarrow \\ & & & & 0 \end{array}$$

If φ' exists we define $\alpha = \kappa_*\tilde{\varphi}'$ and $\tilde{\varphi}\mathcal{J} \subset \varphi_*\mathcal{I}$ follows.

Conversely, assume this condition is satisfied. We first prove

$$\varphi|X'| \subset |Y'|.$$

If $q \notin |Y'|$, then $1 \in \mathcal{J}_q$ and our assumption implies $1 \in (\varphi_*\mathcal{I})_q$. Hence $X' \cap \varphi^{-1}(q) = \emptyset$.

We define $|\varphi'|$ as the restriction of $|\varphi|$ to $|X'|$. The assumption implies the existence of α as above and we may define

$$\tilde{\varphi}' := \kappa^* \alpha .$$

Example. Take $X = \mathbb{C}^2$ with coordinate functions x,y and $Y = \mathbb{C}$ with coordinate function z. Define

$$\varphi : \mathbb{C}^2 \to \mathbb{C}, \ (x,y) \mapsto x + y .$$

Then $\tilde{\varphi}(z) = x + y$. Let $X' \hookrightarrow X$ be the subspace defined by x^2, y^2 and let $Y' \hookrightarrow Y$ be the subspace defined by z^2. Since

$$\tilde{\varphi}(z^2) = x^2 + 2xy + y^2$$

is not a section in the ideal sheaf of X', there is no restriction

$$\varphi' : X' \to Y' .$$

Corollary. If $\varphi : X \to Y$ is a holomorphic map, then there is a uniquely determined holomorphic map $\varphi_{red} : X_{red} \to Y_{red}$ such that the diagram

$$\begin{array}{ccc} X & \xrightarrow{\varphi} & Y \\ \uparrow & & \uparrow \\ X_{red} & \xrightarrow{\varphi_{red}} & Y_{red} \end{array}$$

commutes.

This is clear since $\tilde{\varphi} N_Y \subset \varphi_* N_X$.

0.18. Recall that a local $\mathbb{C}$-algebra R is called *analytic*, if there is an ideal $\mathfrak{a}$ in some power series ring $\mathbb{C}[\langle z_1,\ldots,z_n\rangle]$ such that R is isomorphic (as local $\mathbb{C}$-algebra) to $\mathbb{C}[\langle z_1,\ldots,z_n\rangle]/\mathfrak{a}$.

Lemma. Let R be an analytic local $\mathbb{C}$-algebra with maximal ideal m. Then for $a_1,\ldots,a_n \in m$ there is one and only one homomorphism of local $\mathbb{C}$-algebras

$$\Phi : \mathbb{C}[\langle w_1,\ldots,w_m\rangle] \to R$$

such that $\Phi(w_i) = a_i$ for $i = 1,\ldots,m$.

For a proof see e.g. [50].

0.19. If X is a complex space we denote by $\mathrm{Hol}(X,\mathbb{C}^n)$ the set of holomorphic mappings $X \to \mathbb{C}^n$ and by $z_1,\ldots,z_n \in \mathcal{O}_n(\mathbb{C}^n)$ the coordinate functions.

Proposition. If X is a complex space and $U \subset X$ is open, the map

$$\mathrm{Hol}(U,\mathbb{C}^n) \longrightarrow \mathcal{O}_X(U)^n,$$
$$\varphi \mapsto (\widetilde{\varphi}z_1,\ldots,\widetilde{\varphi}z_n)$$

is bijective.

Proof. In order to show that the map is injective, let $\varphi,\psi \in \mathrm{Hol}(U,\mathbb{C}^n)$ with $\widetilde{\varphi}z_i = \widetilde{\psi}z_i$ be given. For $p \in U$ we have

$$z_i(\varphi(p)) = (\widetilde{\varphi}z_i)(p) = (\widetilde{\psi}z_i)(p) = z_i(\psi(p)),$$

hence $|\varphi| = |\psi|$. Put $q := \varphi(p) = \psi(p)$; we may assume $q = o$. Consider the homomorphisms

$$\widetilde{\varphi}_p,\widetilde{\psi}_p\colon \mathbb{C}[\langle z_1,\ldots,z_n\rangle] = \mathcal{O}_{n,o} \to \mathcal{O}_{X,p}.$$

Since $\widetilde{\varphi}_p(z_i) = \widetilde{\psi}_p(z_i)$, Lemma 0.18 implies $\widetilde{\varphi}_p = \widetilde{\psi}_p$. Hence $\widetilde{\varphi} = \widetilde{\psi}$ and $\varphi = \psi$.

Next we construct an inverse map

$$\mathcal{O}_X(U)^n \longrightarrow \mathrm{Hol}(U,\mathbb{C}^n),$$
$$(f_1,\ldots,f_n) \mapsto \quad \varphi.$$

We may assume U is a local model

$$U \hookrightarrow W \subset \mathbb{C}^m$$

and $f_1,\ldots,f_n$ are induced by $F_1,\ldots,F_n \in \mathcal{O}_m(W)$. Consider the holomorphic map

$$F\colon W \to \mathbb{C}^n,\ p \mapsto (F_1(p),\ldots,F_n(p))$$

and its restriction

$$\varphi\colon U \to \mathbb{C}^n.$$

Then $\widetilde{\varphi}z_1 = f_1,\ldots,\widetilde{\varphi}z_n = f_n$; hence our given map is surjective.

0.20. Corollary. Let X be a complex space. Then the following conditions are equivalent:

i) X is reduced

ii) For any $U \subset X$ open, any complex space Y and any holomorphic maps $\varphi,\psi\colon U \to Y$ the equality $|\varphi| = |\psi|$ implies $\varphi = \psi$.

Proof. i) $\Rightarrow$ ii). The question being local we may assume $Y = \mathbb{C}^n$ and the assertion follows from Lemma 0.16 and 0.19.

ii) $\Rightarrow$ i). If X is not reduced, there is $U \subset X$ and a nilpotent $f \in \mathcal{O}_X(U)$, hence a holomorphic map $f\colon U \to \mathbb{C}$ with $f \neq 0$ but $|f| = 0$.

It should be noted that ii) $\Rightarrow$ i) need not hold if only $U = X$ is admitted in ii).

0.21. If X is a complex space and $p \in X$, the pair (X,p) is called a *germ of a complex space. Morphisms of germs* $(X,p) \to (Y,q)$ are defined as germs of holomorphic maps. If U is an open neighbourhood of p and $\varphi: U \to Y$ is a holomorphic map with $\varphi(p) = q$, we denote by

$$\varphi_p: (X,p) \to (Y,q)$$

the induced germ. Now

$$(X,p) \to \mathcal{O}_{X,p}$$

becomes a contravariant functor from the category of germs of complex spaces to the category of analytic $\mathbb{C}$-algebras in the following way. If a morphism

$$\varphi_p: (X,p) \to (Y,q)$$

is represented by a holomorphic map $\varphi: U \to Y$ where U is an open neighbourhood of p in X, then

$$\tilde{\varphi}_p: \mathcal{O}_{Y,q} \to \mathcal{O}_{X,p}$$

is defined as the associated homomorphism (see 0.3).

Proposition. The functor defined above is an antiequivalence (see [52]).

Proof. We have to verify the following statements.

a) If $\varphi_p, \psi_p: (X,p) \to (Y,q)$ are two morphisms of germs, then $\tilde{\varphi}_p = \tilde{\psi}_p$ implies $\varphi_p = \psi_p$.

b) If R is any local analytic $\mathbb{C}$-algebra, then there exists a germ (X,p) and an isomorphism $\mathcal{O}_{X,p} \to R$.

c) If $\theta: \mathcal{O}_{Y,q} \to \mathcal{O}_{X,p}$ is a homomorphism of $\mathbb{C}$-algebras, then there is a morphism $\varphi_p: (X,p) \to (Y,q)$ such that $\theta = \tilde{\varphi}_p$.

a) Take an open neighbourhood U of p and holomorphic maps

$$\varphi, \psi: U \to Y$$

representing φ_p and ψ_p. The question being local we may assume Y is a local model and it is sufficient to consider the case $Y = \mathbb{C}^n$ and $q = o$. By 0.19 we have

$$\varphi = (\varphi_1, \dots, \varphi_n) \text{ and } \psi = (\psi_1, \dots, \psi_n)$$

with $\varphi_i, \psi_i \in \mathcal{O}_X(U)$. If we consider the homomorphisms

$$\tilde{\varphi}_p, \tilde{\psi}_p: \mathbb{C}[\langle z_1, \dots, z_n \rangle] \to \mathcal{O}_{X,p},$$

then $\widetilde{\varphi}_p = \widetilde{\psi}_p$ implies

$$\varphi_{1,p} = \psi_{1,p}, \ldots, \varphi_{n,p} = \psi_{n,p} \text{ and hence } \varphi_1 = \psi_1, \ldots, \varphi_n = \psi_n$$

if we shrink U sufficiently. 0.19 implies $\varphi = \psi$ and consequently $\varphi_p = \psi_p$.

b) Let $R = \mathbb{C}[\langle w_1, \ldots, w_m \rangle]/\mathfrak{a}$, where $\mathfrak{a} = (f_1, \ldots, f_r)$ and take an open neighbourhood W of o in $\mathbb{C}^m$ with $F_1, \ldots, F_r \in \mathcal{O}_m(W)$ representing $f_1, \ldots, f_r$. Denote by $\mathcal{I} \subset \mathcal{O}_W$ the sheaf of ideals generated by $F_1, \ldots, F_r$ and by $X \hookrightarrow W$ the corresponding complex subspace. Then $\mathcal{O}_{X,o}$ is isomorphic to R (see 0.8 and 0.9).

c) Let $\theta: \mathcal{O}_{Y,q} \to \mathcal{O}_{X,p}$ be given. We may assume $Y \hookrightarrow Z \subset \mathbb{C}^n$ is a local model with q = o and we consider the commutative diagram

$$\begin{array}{ccc}
 & & 0 \\
 & & \downarrow \\
 & & \mathcal{J} \\
 & & \downarrow \\
 & & \mathbb{C}[\langle z_1, \ldots, z_n \rangle] \\
 & \swarrow^{\Theta} & \downarrow \\
\mathcal{O}_{X,p} & \xleftarrow{\ \theta\ } & \mathcal{O}_{Y,o} \\
 & & \downarrow \\
 & & 0 ,
\end{array}$$

where Θ is defined by means of 0.18. By 0.19 there is an open neighbourhood U of p and a holomorphic map

$$\Phi: U \to \mathbb{C}^n \text{ with } \Phi(p) = o \text{ and } \widetilde{\Phi}_p = \Theta .$$

Since $\widetilde{\Phi}_p(\mathcal{J}) = \Theta(\mathcal{J}) = 0$, we may apply 0.17 and by shrinking U we obtain a holomorphic map

$$\varphi: U \to Y \text{ with } \widetilde{\varphi}_p = \theta.$$

It should be noted that we have made use of the Coherence Theorem of OKA (0.12) in this proof.

<u>0.22.</u> The duality statement of 0.21 admits to derive local "geometric" properties of complex spaces and holomorphic maps from algebraic considerations on the local algebras. We mention two immediate consequences.

<u>Corollary 1.</u> Let X be a complex space, $p \in X$. Then the following conditions are equivalent:

i) X is non-singular at p.

ii) The local ring $\mathcal{O}_{X,p}$ is regular.

<u>Proof.</u> i) ⇒ ii) is obvious.

ii) ⇒ i). If $\mathcal{O}_{X,p}$ is regular then there is an isomorphism

$$\theta\colon \mathbb{C}[\langle z_1,\dots,z_n\rangle] \to \mathcal{O}_{X,p}$$

which, by means of 0.21, yields a biholomorphic map

$$\varphi\colon U \to Z$$

between open neighbourhoods U of p in X and Z of o in $\mathbb{C}^n$.

<u>Corollary 2.</u> Let $W \subset \mathbb{C}^m$ and $Z \subset \mathbb{C}^n$ be open, $X \hookrightarrow W$ and $Y \hookrightarrow Z$ be closed complex subspaces and assume a holomorphic map $\varphi\colon X \to Y$ is given. Then for any $p \in X$ and $q := \varphi(p)$ there exists an open neighbourhood W' of p in W together with a holomorphic map $\Phi\colon W' \to Z$ such that the diagram

$$\begin{array}{ccc} W' & \xrightarrow{\ \Phi\ } & Z \\ \uparrow & & \uparrow \\ X\cap W' & \xrightarrow{\ \varphi\ } & Y \end{array}$$

commutes.

We say that Φ *generates* φ.

<u>Proof.</u> We may assume $p = q = o$ and consider the diagram

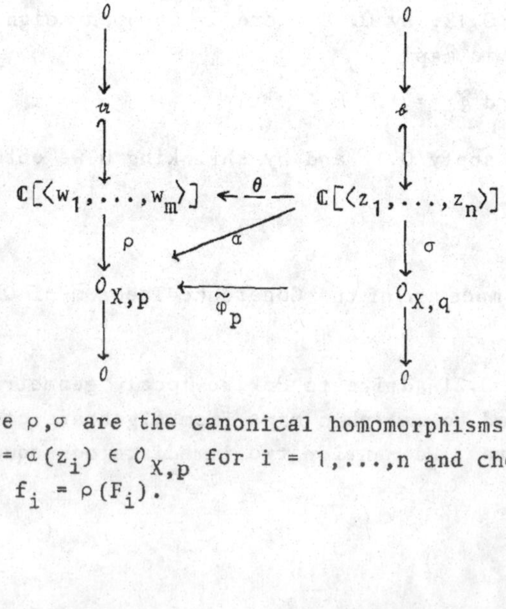

where ρ,σ are the canonical homomorphisms and $\alpha := \tilde{\varphi}_p \circ \sigma$. Define $f_i := \alpha(z_i) \in \mathcal{O}_{X,p}$ for $i = 1,\dots,n$ and choose $F_i \in \mathbb{C}[\langle w_1,\dots,w_m\rangle]$ such that $f_i = \rho(F_i)$.

By 0.18 there is a homomorphism

$$\theta: \mathbb{C}[\langle z_1,\ldots,z_n\rangle] \to \mathbb{C}[\langle w_1,\ldots,w_m\rangle]$$

with $\theta(z_i) = F_i$. By 0.21 there is an open neighbourhood W' of o in W and a holomorphic map

$$\Phi: W' \to \mathbb{C}^n$$

such that $\Phi_o = \theta$. Since θ induces $\tilde{\varphi}_p$, Φ generates φ.

0.23. A holomorphic map φ: X → Y is called an *embedding* if there is a closed subspace ι: Y' ↪ Y and a factorization

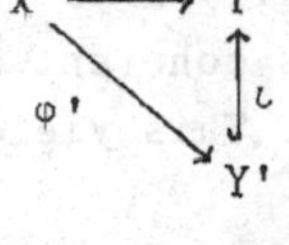

where φ' is biholomorphic.

φ: X → Y is called an *immersion* (resp. *biholomorphic*) *at* p ∈ X if there are open neighbourhoods U of p in X and V of $\varphi(p)$ in Y such that $\varphi|U$: U → V is an embedding (resp. biholomorphic).

φ is called an *immersion* if it is an immersion at every point. φ is called *locally biholomorphic*, if it is biholomorphic at every point.

Proposition. Let φ: X → Y be a holomorphic map.

1) φ is biholomorphic (resp. an immersion) at p ∈ X, if and only if

$$\tilde{\varphi}_p: \mathcal{O}_{Y,\varphi(p)} \to \mathcal{O}_{X,p}$$

is bijective (resp. surjective).

2) φ is an embedding if and only if

a) φ is an immersion

b) $|\varphi|$ is closed and injective (consequently A := $|\varphi|(X) \subset |Y|$ is closed and $|\varphi|: |X| \to A$ is a homeomorphism, i.e. $|\varphi|$ is a *topological embedding*).

In this case $\mathcal{K}er\,\tilde{\varphi} \subset \mathcal{O}_Y$ is the coherent ideal defining Y'.

Proof. 1) For "biholomorphic at p" the assertion is a direct consequence of 0.21.

For an immersion the necessity of the condition is clear. To prove the sufficiency we put q := $\varphi(p)$ and denote by $k \subset \mathcal{O}_{Y,q}$ the kernel of $\tilde{\varphi}_p$. Since $\mathcal{O}_{Y,q}$ is noetherian we may apply 0.9 and find an open neighbourhood V of q in Y and a coherent ideal $\mathcal{K} \subset \mathcal{O}_V$ with $\mathcal{K}_q = k$. Let V' be the closed complex subspace of V defined by K. If U and V are suitably chosen we have a factorization

$$U \xrightarrow{\varphi'} V' \hookrightarrow V$$

of $\varphi|U$. Since $\tilde{\varphi}'_q$ is bijective, φ' is biholomorphic if we shrink U and V.

2) Clearly conditions a) and b) are necessary. To prove their sufficiency we put $K := \operatorname{Ker} \tilde{\varphi}$. Condition 1) implies

$$A = \operatorname{supp}(\mathcal{O}_Y/K), \quad (\varphi_* \mathcal{O}_X)_{\varphi(p)} = \mathcal{O}_{X,p} \text{ and } K_p = \operatorname{Ker} \tilde{\varphi}_p$$

for every $p \in X$.

From the definition of an immersion it follows that locally K is the sheaf of ideals of a closed complex subspace. Hence K is coherent and defines a closed complex subspace Y' of Y with $|Y'| = A$. This yields a factorization

$$X \xrightarrow{\varphi'} Y' \hookrightarrow Y$$

of φ, where φ' is locally biholomorphic by 1). But $|\varphi|$ was injective, hence φ' is biholomorphic.

<u>0.24.</u> Let $(X_i)_{i\in I}$ be a family of complex spaces. Assume that for $i \in I$ a family $(X_{ij})_{j\in I}$ of open subspaces $X_{ij} \subset X_i$ and for $i,j \in I$ a biholomorphic map $\varphi_{ji}: X_{ij} \to X_{ji}$ is given. Then the pair

$$((X_{ij})_{i,j\in I}, (\varphi_{ij})_{i,j\in I})$$

is called *gluing data (for a complex space)* if the following conditions are satisfied:

a) $X_{ii} = X_i$ and $\varphi_{ii} = id_{X_i}$ for every $i \in I$.

b) For every triple $i,j,k \in I$ we have

$$\varphi_{ji}(X_{ij} \cap X_{ik}) \subset X_{ji} \cap X_{jk} \text{ and}$$

$$\varphi_{ki}|(X_{ij} \cap X_{ik}) = (\varphi_{kj}|(X_{jk} \cap X_{ji}))\circ(\varphi_{ji}|(X_{ij} \cap X_{ik})).$$

Let M denote the disjoint union of the topological spaces $|X_i|$ and call $p,q \in M$ equivalent, if there are indices $i,j \in I$ such that $p \in X_i$, $q \in X_j$ and $q = \varphi_{ji}(p)$. Let $|X|$ be the quotient of M by this equivalence relation, provided with the quotient topology. Our gluing data is called *hausdorff*, if $|X|$ is hausdorff.

<u>Proposition.</u> Let a family $(X_i)_{i\in I}$ of complex spaces with hausdorff gluing data $((X_{ij}), (\varphi_{ij}))$ be given. Then there is a complex space X with an open covering $(U_i)_{i\in I}$ and a family of biholomorphic maps $\varphi_i: X_i \to U_i$ such that for $i,j \in I$ the diagram

$$\begin{array}{ccc} X_i \supset X_{ij} & \xrightarrow{\varphi_{ji}} & X_{ji} \subset X_j \\ \downarrow \varphi_i & & \downarrow \varphi_j \\ U_i \supset & U_i \cap U_j & \subset U_j \end{array}$$

commutes.

Proof. Let $\rho_i\colon |X_i| \to |X|$ be the canonical continuous map and put $U_i := \rho_i(X_i)$. In order to construct $\mathcal{O}_X$ we first define a sheaf $\mathcal{O}_i$ on U_i by

$$\mathcal{O}_i(U) := \mathcal{O}_{X_i}(\rho_i^{-1}(U))$$

for $U \subset U_i$ open. The biholomorphic maps φ_{ji} determine isomorphisms

$$\theta_{ji}\colon \mathcal{O}_i|(U_i \cap U_j) \to \mathcal{O}_j|(U_i \cap U_j).$$

Hence we may define

$$\mathcal{O}_X(U) := \{(f_i) \in \prod_i \mathcal{O}_i(U \cap U_i):$$

$$\theta_{ji}(f_i)|(U_i \cap U_j \cap U) = f_j|(U_i \cap U_j \cap U)\}$$

for $U \subset |X|$ open, and $X = (|X|, \mathcal{O}_X)$ has the desired properties.

0.25. Let complex spaces X_1, X_2 and Y together with holomorphic maps $\varphi_i\colon X_i \to Y$ $(i = 1,2)$ be given. A complex space $X_1 \times_Y X_2$ together with holomorphic maps

$$\pi_i\colon X_1 \times_Y X_2 \to X_i$$

such that $\varphi_1 \circ \pi_1 = \varphi_2 \circ \pi_2$ is called *fibre product* of X_1 and X_2 over Y (or more precisely over φ_1 and φ_2) if it has the following universal property: Given any complex space X together with holomorphic maps $\psi_i\colon X \to X_i$ such that $\varphi_1 \cdot \psi_1 = \varphi_2 \cdot \psi_2$, there is a unique holomorphic map $\psi\colon X \to X_1 \times_Y X_2$ such that the diagram

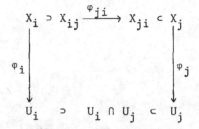

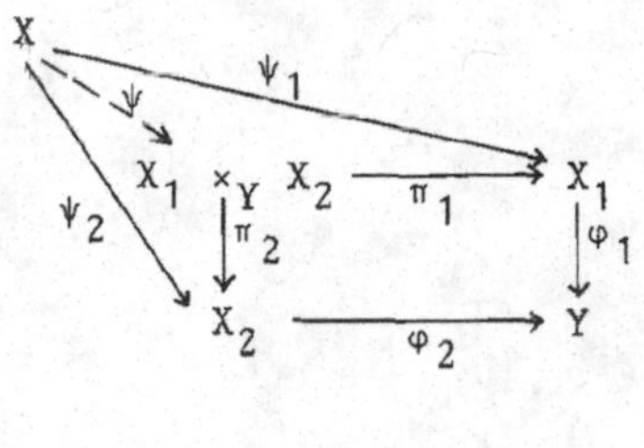

commutes.

A commutative diagram

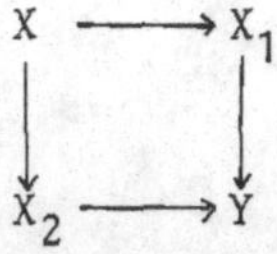

of holomorphic maps is called a *cartesian square* if X is a fibre product of X_1 and X_2 over Y.

In case Y is a simple point, a fibre product of X_1 and X_2 is called *direct product* and denoted by $X_1 \times X_2$. Its universal property then is the following: Given any complex space X together with holomorphic maps $\psi_i: X \to X_i$ there is a unique holomorphic map $(\psi_1, \psi_2): X \to X_1 \times X_2$ such that the diagram

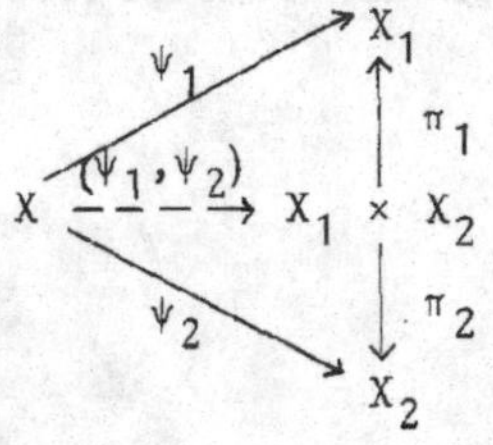

commutes.

The proof of the existence of a fibre product will be given in several steps in the following sections. We recall that the universal property implies the uniqueness up to isomorphisms.

<u>0.26. Proposition.</u> $\mathbb{C}^{m+n}$ together with the canonical projections

$$\pi_1: \mathbb{C}^{m+n} \to \mathbb{C}^m \text{ and}$$

$$\pi_2: \mathbb{C}^{m+n} \to \mathbb{C}^n$$

is a direct product of $\mathbb{C}^m$ and $\mathbb{C}^n$.

<u>Proof.</u> If $\psi_1: X \to \mathbb{C}^m$ and $\psi_2: X \to \mathbb{C}^n$ are given, then $\psi_1 = (f_1, \ldots, f_m)$ and $\psi_2 = (g_1, \ldots, g_n)$ with $f_1, \ldots, f_m, g_1, \ldots, g_n \in \mathcal{O}_X(X)$ by 0.19, and

$$(\psi_1, \psi_2) := (f_1, \ldots, f_m, g_1, \ldots, g_n)$$

is as desired.

0.27. Proposition. Let $\varphi: X \to Y$ be a holomorphic map and Y' an open (resp. closed) complex subspace of Y. Then there is a cartesian square

$$\begin{array}{ccc} \varphi^{-1}(Y') & \xrightarrow{\varkappa} & X \\ \downarrow{\scriptstyle \varphi'} & & \downarrow{\scriptstyle \varphi} \\ Y' & \xrightarrow{\iota} & Y, \end{array}$$

where $\varphi^{-1}(Y')$ is an open (resp. closed) complex subspace of X and $\varkappa, \iota$ denote the canonical injections. In particular there is an isomorphism

$$\mathcal{O}_{\varphi^{-1}(Y')} \longrightarrow (\varphi')^*\mathcal{O}_{Y'} \ . \qquad (*)$$

$\varphi^{-1}(Y)$ is called the *inverse image* of Y'.

In particular, if $Y' = (\{q\},\mathbb{C})$ is a simple point, then $\varphi^{-1}(Y')$ is called the *fibre* of φ over q and denoted by X_q .

Proof. In case Y' is open we define $\varphi^{-1}(Y')$ as the open subspace with $\varphi^{-1}(|Y'|)$ as underlying topological space and the assertion is obvious. If $Y' \hookrightarrow Y$ is a closed complex subspace generated by the ideal $\mathcal{J} \subset \mathcal{O}_Y$ we consider the exact sequences

$$0 \longrightarrow \mathcal{J} \longrightarrow \mathcal{O}_Y \longrightarrow \mathcal{O}_Y/\mathcal{J} \longrightarrow 0 \quad \text{and}$$

$$\varphi^*\mathcal{J} \xrightarrow{\mu} \mathcal{O}_X \longrightarrow \varphi^*(\mathcal{O}_Y/\mathcal{J}) \longrightarrow 0 \qquad (**)$$

(see 0.10). Define $\mathcal{I} \subset \mathcal{O}_X$ as the image of the multiplication map μ and $\varphi^{-1}(Y') \hookrightarrow X$ as the closed complex subspace generated by $\mathcal{I}$. More explicitely $\mathcal{I}$ can be described in the following way: if $q \in Y$ and $\mathcal{J}_q$ is generated by $a_1,..,a_k \in \mathcal{O}_{Y,q}$ then for every $p \in X$ with $\varphi(p) = q$ the stalk $\mathcal{I}_p$ is generated by $\widetilde{\varphi}_p(a_1),..,\widetilde{\varphi}_p(a_k) \in \mathcal{O}_{X,p}$.

By construction of $\mathcal{I}$, there is a restriction φ' of φ (see 0.17). The exact sequence

$$0 \longrightarrow \mathcal{I} \longrightarrow \mathcal{O}_X \longrightarrow \mathcal{O}_X/\mathcal{I} \longrightarrow 0$$

together with (**) yields (*).

In order to prove the universal property let a commutative diagram

$$\begin{array}{ccccc} S & & \xrightarrow{\lambda} & & X \\ & \searrow{\scriptstyle \psi} & \varphi^{-1}(Y') & \xrightarrow{\varkappa} & X \\ & & \downarrow{\scriptstyle \varphi} & & \downarrow{\scriptstyle \varphi} \\ & & Y' & \xrightarrow{\iota} & Y \end{array}$$

of holomorphic maps be given. 0.17 implies $\widetilde{(\iota\circ\psi)}(\mathcal{J}) = 0$. Since $\mathcal{I}$ was generated by $\varphi^{-1}(\mathcal{J})$ and the diagram commutes we obtain $\widetilde{\lambda}(\mathcal{I}) = 0$. Hence 0.17 implies the existence of a unique holomorphic map $S \to \varphi^{-1}(Y')$.

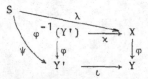

Example. Consider the holomorphic map

$$\varphi: \mathbb{C} \to \mathbb{C},\ x \mapsto x^2.$$

Let $Y' \hookrightarrow \mathbb{C}$ be the origin, i.e. the simple point with the ideal generated by the coordinate function y. Then $\varphi^{-1}(Y')$ is the subspace defined by x^2, i.e. a double point. In particular, $\varphi^{-1}(Y')$ is not reduced. This is a very characteristic example how non reduced complex spaces can occur in a natural way.

Corollary. Let X be a complex space with (open or closed) complex subspaces X_1 and X_2. Then there is a unique complex subspace $X_1 \cap X_2$ of X_1 and X_2 such that the square of canonical injections

$$\begin{array}{ccc} X_1 \cap X_2 & \longrightarrow & X_1 \\ \downarrow & & \downarrow \\ X_2 & \longrightarrow & X \end{array}$$

is cartesian.

$X_1 \cap X_2$ is called the *analytic intersection* of X_1 and X_2.

Obviously $|X_1 \cap X_2| = |X_1| \cap |X_2|$. If X_1 and X_2 are closed and defined by ideals $I_1, I_2 \subset \mathcal{O}_X$ the analytic intersection $X_1 \cap X_2$ is defined by $I_1 + I_2$.

0.28. Proposition. Let complex spaces X and Y with open (resp. closed) complex subspaces X' and Y' be given. If the direct product $X \times Y$ exists, then $X' \times Y'$ exists and is an open (resp. closed) subspace of $X \times Y$.

Proof. Denote by $\pi_1: X \times Y \to X$ and $\pi_2: X \times Y \to Y$ the canonical projections and by $\iota: X' \to X$ and $\kappa: Y' \to Y$ the canonical injections. By 0.27 we have a commutative diagram

$$\begin{array}{ccccc} Z & \longrightarrow & X'' & \longrightarrow & X' \\ \downarrow & & \downarrow & & \downarrow \iota \\ Y'' & \longrightarrow & X \times Y & \xrightarrow{\pi_1} & X \\ \downarrow & & \downarrow \pi_2 & & \\ Y' & \xrightarrow{\kappa} & Y & & \end{array}$$

of holomorphic maps with cartesian squares.

Hence

$$X' \times Y' := Z$$

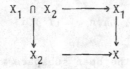

has the desired properties.

Corollary. If $W \subset \mathbb{C}^m$ and $Z \subset \mathbb{C}^n$ are open and $X \hookrightarrow W$, $Y \hookrightarrow Z$ are closed complex subspaces, then $X \times Y$ exists and is a closed complex subspace of $W \times Z$.

In a special case we give a useful and more explicit description of $X \times Y$. Let X and Y be defined by coherent ideals $I_X \subset O_W$ and $I_Y \subset O_Z$ and assume I_X resp. I_Y are generated by $f_1, \ldots, f_k \in O_W(W)$ resp. $g_1, \ldots, g_l \in O_Z(Z)$.

The open subspace $W \times Z \subset \mathbb{C}^{m+n}$ with the canonical projections $\pi_1 \colon W \times Z \to W$ and $\pi_2 \colon W \times Z \to Z$ is a direct product. Put

$$f_i \otimes 1 := \tilde{\pi}_1(f_i) \in O_{W\times Z}(W\times Z) \quad \text{and}$$

$$1 \otimes g_i := \tilde{\pi}_2(g_j) \in O_{W\times Z}(W\times Z)$$

for $i = 1, \ldots, k$, and $j = 1, \ldots, l$ and let $J \subset O_{W\times Z}$ be the ideal generated by

$$f_1 \otimes 1, \ldots, f_k \otimes 1, \ 1 \otimes g_1, \ldots, 1 \otimes g_l .$$

Then $X \times Y$ is the subspace of $W \times Z$ defined by J and for $p \in X$, $q \in Y$ we have

$$O_{X\times Y,(p,q)} = O_{X,p} \mathbin{\hat{\otimes}} O_{Y,q}$$

(for details about the "analytic tensor product" we refer to [105], [50]).

0.29. Proposition. If X and Y are complex spaces then the direct product $X \times Y$ exists.

In particular, if $\varphi_i \colon X_i \to Y_i$ $(i = 1,2)$ are holomorphic maps, there is a uniquely determined holomorphic map $\varphi_1 \times \varphi_2$ such that the diagram

$$\begin{array}{ccc} X_1 & \xrightarrow{\varphi_1} & Y_1 \\ \uparrow & & \uparrow \\ X_1 \times X_2 & \xrightarrow{\varphi_1 \times \varphi_2} & Y_1 \times Y_2 \\ \downarrow & & \downarrow \\ X_2 & \xrightarrow{\varphi_2} & Y_2 \end{array}$$

commutes. (The vertical arrows denote the canonical projections.)

Proof. We use the local solution of 0.28 and the gluing procedure of 0.24.

Let

$$X = \bigcup_{k\in K} X_k \quad \text{and} \quad Y = \bigcup_{l\in L} Y_l$$

be open coverings by local models. Put $I = K \times L$ and $S_i = X_k \times Y_l$ for $i = (k,l)$. If $j = (k',l') \in I$ then

$$(X_k \cap X_{k'}) \times (Y_l \cap Y_{l'})$$

is isomorphic to an open complex subspace of $S_i = X_k \times Y_l$ which we denote by S_{ij}. By symmetry it is also isomorphic to an open complex subspace S_{ji} of $S_j = X_{k'} \times Y_{l'}$. This yields a canonical biholomorphic map

$$\varphi_{ji}: S_{ij} \to S_{ji}.$$

Obviously $((S_{ij}), (\varphi_{ji}))$ is a gluing data. It is hausdorff since the obtained topological space is $|X| \times |Y|$. Now Proposition 0.24 yields $X \times Y$ and the projections of $X_k \times Y_l$ determine projections of $X \times Y$ on its factors.

<u>0.30.</u> In order to obtain the fibre product from the ordinary product of complex spaces we need the so called "diagonal". We start with the algebraic

<u>Lemma.</u> Let R be a local analytic algebra with maximal ideal m and denote by

$$\mu: R \hat{\otimes} R \to R$$

the homomorphism induced from the multiplication in R. Then

$$\operatorname{Ker} \mu = \{a\otimes 1 - 1\otimes a : a \in m\}.$$

We only sketch the <u>proof.</u> In case

$$R = \mathbb{C}[\langle x_1,\ldots,x_n\rangle],$$

$$R \hat{\otimes} R = \mathbb{C}[\langle x_1',\ldots,x_n',x_1'',\ldots,x_n''\rangle] \text{ with } x_i' = x_i\otimes 1 \text{ and } x_i'' = 1\otimes x_i$$

for $i = 1,\ldots,n$. The homomorphism μ is given by

$$\mathbb{C}[\langle x',x''\rangle] \to \mathbb{C}[\langle x\rangle],\ F(x',x'') \mapsto F(x,x).$$

An elementary computation shows that the kernel of μ is generated by

$$x_1' - x_1'',\ldots,x_n' - x_n''.$$

The general case follows by considering the commutative diagram

$$\begin{array}{ccc} \mathbb{C}[\langle x',x''\rangle] & \longrightarrow & \mathbb{C}[\langle x\rangle] \\ \downarrow \rho\hat{\otimes}\rho & & \downarrow \rho \\ R \hat{\otimes} R & \longrightarrow & R \end{array}$$

and using the fact

$$\operatorname{Ker} \rho\hat{\otimes}\rho = \mathfrak{n}\otimes 1 + 1\otimes\mathfrak{n}, \text{ if } \mathfrak{n} := \operatorname{Ker} \rho$$

(for details see [50]).

If X is a complex space, then 0.29 implies the existence of a holomorphic map

$$\delta_X := (\mathrm{id}_X, \mathrm{id}_X) : X \to X \times X,$$

which we call *diagonal map*. It obviously has the following property: If S is any complex space with holomorphic maps

$$\varphi_1 : S \to X \text{ and } \varphi_2 : S \to X,$$

then there is a factorization

$$S \to X \xrightarrow{\delta_X} X \times X$$

of (φ_1, φ_2) if and only if $\varphi_1 = \varphi_2$.

Proposition. For any complex space X the diagonal map δ_X is an embedding, i.e. there is a closed complex subspace $D_X \hookrightarrow X \times X$ and a factorization

$$X \xrightarrow{\delta'_X} D_X \hookrightarrow X \times X$$

of δ_X, where δ'_X is biholomorphic.

D_X is called the *diagonal* in $X \times X$.

If $p \in X$ and $f_1, \ldots, f_k$ are holomorphic in a neighbourhood of p such that $f_{1,p}, \ldots, f_{k,p}$ generate the maximal ideal $m_{X,p}$, then

$$f_1\otimes 1 - 1\otimes f_1, \ldots, f_k\otimes 1 - 1\otimes f_k$$

generate the sheaf of ideals of D_X at $(p,p) \in X \times X$.

Proof. Since $|X|$ is hausdorff, $|\delta_X|$ is a topological embedding. For $p \in X$ we consider the canonical ring homomorphism

$$\tilde{\delta}_{X,p} : \mathcal{O}_{X\times X,(p,p)} = \mathcal{O}_{X,p} \hat{\otimes} \mathcal{O}_{X,p} \to \mathcal{O}_{X,p}.$$

By construction of δ_X we have

$$\tilde{\delta}_{X,p}(f\otimes 1) = f \text{ and } \tilde{\delta}_{X,p}(1\otimes f) = f \qquad (*)$$

for every $f \in \mathcal{O}_{X,p}$. Hence $\tilde{\delta}_{X,p}$ is obviously surjective, and Proposition 0.23 implies that δ_X is an embedding. Furthermore $(*)$ implies that $\tilde{\delta}_{X,p}$ is induced by the multiplication, hence the lemma implies the second assertion.

We still mention a very explicit description of the diagonal.

First we consider the complex space $\mathbb{C}^n$ with coordinate functions $x_1,\ldots,x_n$ and by $D_n \hookrightarrow \mathbb{C}^n \times \mathbb{C}^n$ we denote its diagonal. Obviously

$$|D_n| = \{(p,q) \in \mathbb{C}^n \times \mathbb{C}^n : p = q\}.$$

We denote by

$$x_1\otimes 1,\ldots,x_n\otimes 1,\ 1\otimes x_1,\ldots,1\otimes x_n$$

the coordinate functions in $\mathbb{C}^n \times \mathbb{C}^n$ and we claim that

$$x_1\otimes 1 - 1\otimes x_1,\ldots,x_n\otimes 1 - 1\otimes x_n$$

are global generators of the sheaf of ideals defining D_n in $\mathbb{C}^n \times \mathbb{C}^n$. But if $p = (p_1,\ldots,p_n) \in \mathbb{C}^n$ is any point, then

$$x_1-p_1,\ldots,x_n-p_n \in m_{n,p} \subset \mathcal{O}_{n,p}$$

are generators of this maximal ideal. We also have

$$(x_i-p_i)\otimes 1 - 1\otimes(x_i-p_i) = x_i\otimes 1 - 1\otimes x_i$$

for $i = 1,\ldots,n$, hence our assertion follows from the lemma.

If $W \subset \mathbb{C}^n$ is an open subspace, then

$$D_W \hookrightarrow W \times W$$

is the subspace generated by the restrictions of

$$x_1\otimes 1 - 1\otimes x_1,\ldots,x_n\otimes 1 - 1\otimes x_n$$

to $W \times W$ for the same reasons.

Now, if $X \hookrightarrow W$ be a closed complex subspace generated by $g_1,\ldots,g_r \in \mathcal{O}_n(W)$, we consider the commutative diagram

$$\begin{array}{ccccc} X & \xrightarrow{\delta_X'} & D_X & \hookrightarrow & X \times X \\ \downarrow & & \downarrow & & \downarrow \\ W & \xrightarrow{\delta_W'} & D_W & \hookrightarrow & W \times W. \end{array}$$

Obviously the right square is cartesian, i.e. D_X is the analytic intersection of D_W and $X \times X$ (see 0.27). Since $X \times X \hookrightarrow W \times W$ is the subspace generated by

$$g_1 \otimes 1, \dots, g_r \otimes 1,\ 1 \otimes g_1, \dots, 1 \otimes g_r,$$

$D_X \hookrightarrow W \times W$ is the subspace generated by

$$x_i \otimes 1 - 1 \otimes x_i,\ g_j \otimes 1,\ 1 \otimes g_j,\ i = 1,\dots,n \text{ and } j = 1,\dots,r.$$

<u>0.31. Corollary.</u> Let complex spaces S,X with holomorphic maps $\varphi_1, \varphi_2: S \to X$ be given. Then there is a factorization

$$\begin{array}{ccc} S & \xrightarrow{\varphi} & X \times X \\ & {}_{\varphi'}\searrow & \uparrow J \\ & & D_X \end{array}$$

of $\varphi = (\varphi_1, \varphi_2)$ if and only if $\varphi_1 = \varphi_2$.

In particular, if ρ_i $(i = 1,2)$ denotes the restriction of the canonical projections $\pi_i: X \times X \to X$, then $\rho_1 = \rho_2$ and the square

$$\begin{array}{ccc} D_X & \xrightarrow{\rho} & X \\ \rho \downarrow & & \downarrow id_X \\ X & \xrightarrow[id_X]{} & X \end{array}$$

with $\rho := \rho_1 = \rho_2$ is cartesian.

<u>0.32. Corollary.</u> For complex spaces X_1, X_2, Y and holomorphic maps $\varphi_i: X_i \to Y$ $(i = 1,2)$ the fibre product $X_1 \times_Y X_2$ exists and is a closed complex subspace of $X_1 \times X_2$.

For any $p \in X_2$ the fibre $(X_1 \times_Y X_2)_p$ over p is isomorphic to the fibre $(X_1)_{\varphi_2(p)}$

<u>Proof.</u> Consider the cartesian square

$$\begin{array}{ccc} X_1 \times_Y X_2 & \hookrightarrow & X_1 \times X_2 \\ \downarrow & & \downarrow \varphi_1 \times \varphi_2 \\ D_Y & \hookrightarrow & Y \times Y \end{array}$$

and define $\pi_i: X_1 \times_Y X_2 \to X_i$ as the restriction of the canonical projections $X_1 \times X_2 \to X_i$ $(i = 1,2)$. Then the universal property is an easy consequence of 0.31. It immediately implies the assertion on the fibres.

0.33. Proposition. Let complex spaces X, Y and a pair $\varphi_1, \varphi_2: X \to Y$ of holomorphic maps be given. Then there exists a closed complex subspace $\iota: X' \hookrightarrow X$ with $\varphi_1 \circ \iota = \varphi_2 \circ \iota$ such that the following universal property is satisfied:

If $\psi: S \to X$ is any holomorphic map with $\varphi_1 \circ \psi = \varphi_2 \circ \psi$, then there is a uniquely determined holomorphic map $\psi': S \to X'$ such that the diagram

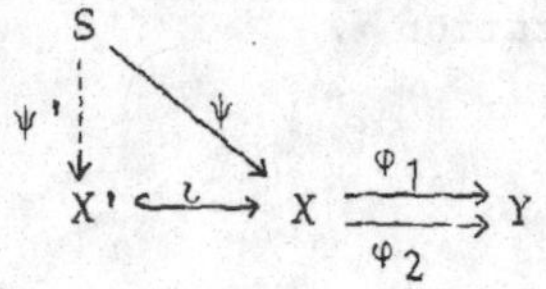

commutes.

X' is called *kernel of the double arrow* $X \overset{\varphi_1}{\underset{\varphi_2}{\rightrightarrows}} Y$.

Proof. Consider the diagram

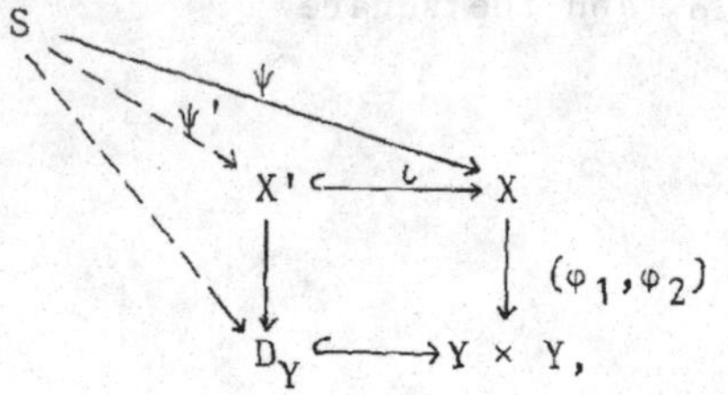

where $X' := (\varphi_1, \varphi_2)^{-1}(D_Y)$. Now $\varphi_1 \circ \psi = \varphi_2 \circ \psi$ means that $(\varphi_1 \circ \psi, \varphi_2 \circ \psi) = (\varphi_1, \varphi_2) \circ \psi$ can be factored through D_Y by 0.31 and the universal property of X' implies the assertion.

0.34. In the category of reduced complex spaces and holomorphic maps there is a universal problem analogous to 0.25.

Proposition. Let X_1, X_2 be reduced complex spaces. Then the direct product $X_1 \times X_2$ in the category of arbitrary complex spaces is reduced and hence a direct product in the category of reduced complex spaces.

If a reduced complex space Y and holomorphic maps $\varphi_i: X_i \to Y$ $(i = 1,2)$ are given, then the subspace

$$(X_1 \times_Y X_2)_{red} \hookrightarrow X_1 \times X_2$$

is a fibre product of X_1 and X_2 over Y in the category of reduced complex spaces.

Note that in general $X_1 \times_Y X_2$ is <u>not</u> reduced if X_1, X_2 and Y are reduced.

<u>Proof.</u> For the first assertion we refer to [50]. The second assertion follows by applying the functor red.

<u>0.35.</u> In the local theory of holomorphic mappings we will frequently use the following

<u>Proposition.</u> Let $\varphi: X \to Y$ be a holomorphic map, $p \in X$, $q := \varphi(p)$. Assume there is a holomorphic map $\psi: X_q \to \mathbb{C}^k$ which is an immersion at p. Then there exists an open neighbourhood U of p in X and a holomorphic map $\chi: U \to Y \times \mathbb{C}^k$ such that

1) χ is an immersion at p
2) the diagram

$$\begin{array}{ccc} X_q \cap U & \xrightarrow{\ \psi\ } & \{q\} \times \mathbb{C}^k \\ \downarrow & & \downarrow \\ U & \xrightarrow{\ \chi\ } & Y \times \mathbb{C}^k \\ & \searrow^{\varphi} & \downarrow \pi_1 \\ & & Y \end{array}$$

commutes [45].

<u>Proof.</u> By using 0.19 we can find U with a holomorphic map $\Psi: U \to \mathbb{C}^k$ generating ψ on $X_q \cap U$. If we define $\chi := (\varphi, \Psi)$ then 2) is obvious. To prove 1), we consider the commutative diagram

$$\begin{array}{ccc} \mathcal{O}_{X_q,p} & \xleftarrow{\ \tilde{\psi}_p\ } & \mathbb{C}[\langle t_1,\dots,t_k\rangle] \\ \tilde{\iota}_p \uparrow & \swarrow^{\tilde{\Psi}_p} & \downarrow \\ \mathcal{O}_{X,p} & \xleftarrow{\ \tilde{\chi}_p\ } & \mathcal{O}_{Y,q} \hat{\otimes} \mathbb{C}[\langle t_1,\dots,t_k\rangle] \\ & \nwarrow^{\tilde{\varphi}_p} & \uparrow \\ & & \mathcal{O}_{Y,q} \end{array}$$

Since $\tilde{\psi}_p(t_1),\dots,\tilde{\psi}_p(t_n)$ generate the maximal ideal of $\mathcal{O}_{X_q,p}$ and since $\tilde{\varphi}_p(m_{Y,q})$ generates the kernel of $\tilde{\iota}_p$ by definition of X_q, we know that $\mathcal{O}_{X,p}$ is generated by the images of $\tilde{\varphi}_p$ and $\tilde{\Psi}_p$. Hence

$$\tilde{\chi}_p = \tilde{\Psi}_p \hat{\otimes} \tilde{\varphi}_p$$

is surjective and the assertion follows by 0.23.

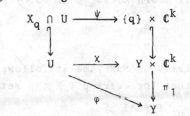

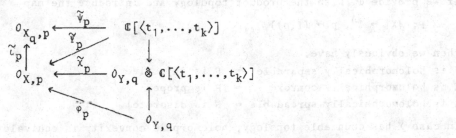

0.36. If X is a complex space we denote by $A := \mathcal{O}_X(X)$ its global function algebra. X is called

holomorphically separable if for any distinct points $p,q \in X$ there is an $f \in A$ with $f(p) \neq f(q)$,

holomorphically convex (resp. *weakly holomorphically convex*) if for any compact $K \subset X$ the set

$$\hat{K} = \{p \in X: |f(p)| \leq \sup|f(K)| \text{ for all } f \in A\}$$

is also compact (resp. has compact connected components),

holomorphically spreadable if every point $p \in X$ has an open neighbourhood U such that

$$\{x \in U: f(x) = f(p) \text{ for all } f \in A\} = \{p\},$$

holomorphically regular if there are *"local coordinates by global functions"*, i.e. for every $p \in X$ the set

$$\{f_p \in \mathcal{O}_{X,p}: f \in A\}$$

generates the maximal ideal $m_{X,p}$. By the Lemma of Nakayama it follows that this is equivalent to the condition that for every $p \in X$ the set

$$\{df(p) := (f_p - f(p)) + m_{X,p}^2 \in m_{X,p}/m_{X,p}^2: f \in A\}$$

generates the $\mathbb{C}$-vector space $m_{X,p}/m_{X,p}^2$.

Clearly a holomorphically separable space is holomorphically spreadable. If we provide $\mathbb{C}^A$ with the product topology and introduce the map

$$F: |X| \to \mathbb{C}^A,\ p \mapsto (f(p))_{f \in A},$$

then we obviously have:

X is holomorphically separable $\Leftrightarrow$ F is injective
X is holomorphically convex $\Leftrightarrow$ F is proper
X is holomorphically spreadable $\Leftrightarrow$ F is discrete

In case X has countable topology, holomorphic convexity is equivalent to the following condition:

If (p_i) is any sequence in X without point of accumulation, then there is an $f \in A$ such that the sequence $(|f(p_i)|)$ is unbounded.

X is called *holomorphically complete* (or *Stein*), if it is holomorphically separable and convex. We state some fundamental results.

0.37. Proposition. A holomorphically spreadable complex space has countable topology [41], [77].

<u>Theorem.</u> Let X be a complex space such that X_{red} is Stein and let F be a coherent O_X-module. Then

A. For every $p \in X$ the set $\{s_p \in F_p : s \in F(X)\}$ generates F_p over $O_{X,p}$.
B. $H^k(X,F) = 0$ for $k \geq 1$.

In case X is reduced this was proved by CARTAN-SERRE [20]. The general case is due to GRAUERT [42]. As usual we refer to these assertions as Theorem A and B.

<u>Corollary 1.</u> A complex space X is Stein if and only if X_{red} is Stein.

<u>Proof.</u> If $N \subset O_X$ denotes the nilradical, then $O_{X_{red}} = O_X/N$ (see 0.16) and we have a canonical homomorphism

$$\rho: O_X(X) \to (O_X/N)(X).$$

Hence clearly X_{red} is Stein if X was Stein. To see the converse we prove that ρ is surjective if X_{red} is Stein.

For i = 1,2,... we consider the exact sequence

$$0 \to N^i/N^{i+1} \to O_X/N^{i+1} \to O_X/N^i \to 0.$$

Since $N \cdot (N^i/N^{i+1}) = 0$, the coherent O_X-module N^i/N^{i+1} is also a coherent O_X/N-module. Since X_{red} is Stein, Theorem B implies

$$H^1(X,N^i/N^{i+1}) = 0 \text{ and } (O_X/N^{i+1})(X) \to (O_X/N^i)(X)$$

is surjective. Hence we have a chain of epimorphisms

$$\varprojlim_i (O_X/N^i)(X) \to . \; . \; . \to (O_X/N^2)(X) \to (O_X/N)(X).$$

Since the projective limit commutes with the section-functor,

$$\varprojlim (O_X/N^i)(X) = (\varprojlim O_X/N^i)(X).$$

The proposition of 0.15 implies

$$\varprojlim O_X/N^i = O_X,$$

hence ρ is surjective.

It should be noted that there are complex spaces which are *not* holomorphically separable (resp. convex) such that the reduction is holomorphically separable (resp. convex) [126], [2]

<u>Corollary 2.</u> A complex space X is Stein if and only if for every exact sequence

$$0 \to F' \to F \to F'' \to 0$$

of coherent $\mathcal{O}_X$-modules the sequence

$$0 \to \mathcal{F}'(X) \to \mathcal{F}(X) \to \mathcal{F}''(X) \to 0$$

is also exact.

Proof. By Theorem B the assertion is necessary. If $p \neq q$ are points of X, denote by $\mathcal{I}$ the sheaf of ideals of the set $\{p,q\}$. We obtain an exact sequence

$$0 \to \mathcal{I}(X) \to \mathcal{O}_X(X) \to (\mathcal{O}_X/\mathcal{I})(X) \to 0,$$

hence X is holomorphically separable. Then X has countable topology (0.35) and holomorphic convexity follows as above, if we take for $\mathcal{I}$ the sheaf of ideals of a sequence without point of accumulation.

0.38. Stein spaces can be characterized in various ways. We state a result of GRAUERT [41, 8].

Theorem. A Stein space is holomorphically regular.

A complex space X is Stein if and only if
a) X is holomorphically spreadable and
b) X is weakly holomorphically convex.

0.39. If X is a complex space with countable topology and $\mathcal{F}$ is a coherent $\mathcal{O}_X$-module, then $\mathcal{F}(X)$ can be provided with the structure of a Frêchet space (see e.g. [51]). If for $x \in X$ the $\mathcal{O}_{X,x}$-module $\mathcal{F}_x$ carries the Krull topology, then the canonical map

$$\rho_x: \mathcal{F}(X) \to \mathcal{F}_x, \; f \mapsto f_x,$$

is continuous. This may be proved by using CAUCHYs integral formula.

Lemma. Let X be a complex space with countable topology, $\mathcal{F}$ a coherent $\mathcal{O}_X$-module, $p \in X$, $G \subset \mathcal{F}_p$ a submodule and $f_1,\dots,f_r \in \mathcal{F}(X)$ such that

$$G = \sum_{i=1}^{r} \mathcal{O}_{X,p}\, f_{i,p}.$$

Then there is a neighbourhood V of p, *privileged* in the following sense: for *every* $f \in \mathcal{F}(X)$ with $f_p \in G$ there are $a_1,\dots,a_r \in \mathcal{O}_X(V)$ such that

$$f = \sum_{i=1}^{r} a_i f_i \text{ on } V.$$

This assertion was first proved by H. CARTAN [18] (see also [57], II D) for domains in $\mathbb{C}^n$, including estimates for $a_1,\dots,a_r$, which we do not need here.

Proof (due to O. FORSTER). By KRULLs intersection theorem $G \subset F_p$ is closed, hence

$$M := \rho_p^{-1}(G) \subset F(X)$$

is closed and a Fréchet space. Denote by $G \subset F$ the subsheaf generated by $f_1,\ldots,f_r$. If $V_o \supset V_1 \supset \ldots$ is a fundamental system of Stein neighbourhoods of p, we define

$$M_j := \{f \in F(X): f_x \in G_x \text{ for every } x \in V_j\} = \bigcap_{x \in V_j} \rho_x^{-1} G_x$$

for $j = 0,1,2,\ldots$. Since $G_x \subset F_x$ is closed and ρ_x is continuous, $M_j \subset M$ is a closed subspace. By Theorem B (0.37) for any $f \in M_j$ we may find $a_1,\ldots,a_r \in O_X(V_j)$ such that $f = a_1 f_1 + \ldots + a_r f_r$ on V_j.

If $f \in M$, then there are $a_{1,p},\ldots,a_{r,p} \in O_{X,p}$ such that

$$f_p = \sum a_{i,p} f_{i,p},$$

hence we may find a neighbourhood U of p and representatives $a_1,\ldots,a_r \in O_X(U)$ such that

$$f = \sum a_i f_i \text{ on } U.$$

Since there is a $n \in \mathbb{N}$ with $V_n \subset U$, we have $f \in M_n$, and hence

$$M = \bigcup_{j=0}^{\infty} M_j.$$

By applying the theorem of BAIRE, we may find $\mu \in \mathbb{N}$ such that $M = M_\mu$, and $V := V_\mu$ is as desired.

0.40. Proposition. Let X be a complex space with countable topology, F a coherent O_X-module, $A \subset F(X)$ an arbitrary subset and $G \subset F$ the subsheaf generated by A. Then G is coherent.

Proof. For any $p \in X$ the submodule $G_p \subset F_p$ is finitely generated. Hence we may find $f_1,\ldots,f_r \in A$ such that

$$G_p = \sum_{i=1}^{r} O_{X,p}\, f_{i,p}.$$

Let $G' \subset F$ be the subsheaf generated by $f_1,\ldots,f_r$. G' is coherent and it is sufficient to verify that there is a neighbourhood V of p in U such that

$$G'|V = G|V.$$

Now clearly $G'_p = G_p$ and $G' \subset G$. Choose V privileged with respect to G_p and $f_1,\ldots,f_r \in F(X)$ (as in 0.39). For any $f \in A$ we obtain $a_1,\ldots,a_r \in \mathcal{O}_X(V)$ such that

$$f = \sum a_i f_i \text{ on } V,$$

and hence $f_x \in G'_x$ for every $x \in V$. This implies

$$G|V \subset G'|V$$

and the assertion is proved.

Our proposition implies a "noetherian" property for coherent sheaves (see [42], [131]).

Corollary. Let F be a coherent sheaf over a complex space X and assume a chain

$$F_0 \subset F_1 \subset F_2 \subset \ldots \subset F$$

of coherent subsheaves is given. Then this chain is stationary over any relatively compact subset of X.

Proof. It is sufficient to show that for every $p \in X$ we can find a neighbourhood V and $n \in \mathbb{N}$ such that

$$F_n|V = F_{n+1}|V = F_{n+2}|V = \ldots \subset F|V.$$

Take a Stein neighbourhood U of p and denote by $G \subset F|U$ the subsheaf generated by

$$\bigcup_{i=0}^{\infty} F_i(U).$$

Our proposition implies that G is coherent. Since F_p is a noetherian module, there is an $n \in \mathbb{N}$ such that

$$F_{n,p} = F_{n+1,p} = F_{n+2,p} = \ldots \subset F_p.$$

By Theorem A (0.37) $F_{i,p}$ is generated by $\{f_p : f \in F_i(U)\}$ for every $i \in \mathbb{N}$. Consequently

$$F_{n,p} = F_{n+1,p} = F_{n+2,p} = \ldots = G_p \subset F_p$$

and this implies (see 0.9) that there is a neighbourhood $V \subset U$ of p such that

$$F_n|V = F_{n+1}|V = F_{n+2}|V = \ldots = G|V.$$

The reader is warned not to make this last conclusion without using the coherent sheaf G.

0.41. We recall some well known algebraic and sheaf theoretic notions. If $P,Q \subset M$ are modules over a ring R, the ideal

$$Q:P = \{a \in R: a\cdot P \subset Q\} \subset R$$

is called the *transporter ideal* of P into Q. In particular

$$\text{Ann } M := 0:M = \{a \in R: a\cdot M = 0\}$$

is called the *annihilator* of M. If for $a \in R$

$$\mu_a: M \to M, \ x \mapsto ax,$$

denotes multiplication by a we consider the canonical homomorphism

$$\varphi: R \to \text{Hom}_R(M,M), \ a \mapsto \mu_a,$$

and obviously

$$\text{Ann } M = \text{Ker } \varphi.$$

Every transporter can be considered as an annihilator since

$$Q:P = \text{Ann}((P+Q)/Q).$$

In the same way *sheaves of transporter and annihilator ideals* can be defined. Here we are only interested in the case where X is a complex space and $P,Q \subset M$ are coherent $\mathcal{O}_X$-modules. If

$$\Phi: \mathcal{O}_X \to \mathcal{H}om_{\mathcal{O}_X}(M,M)$$

denotes the canonical homomorphism, we define

$$\mathcal{A}nn\ M := \mathcal{K}er\ \Phi \text{ and } Q:P := \mathcal{A}nn((P+Q)/Q).$$

Since $\mathcal{H}om_{\mathcal{O}_X}(M,M)$ is coherent, $\mathcal{A}nn\ M$ and $Q:P$ are coherent sheaves of ideals (0.6). Furthermore

$$(\mathcal{A}nn\ M)_p = \text{Ann } M_p \text{ and } (Q:P)_p = Q_p:P_p$$

for every $p \in X$ (0.8). For details we refer to [1].

0.42. Let X be a complex space and suppose $I \subset \mathcal{O}_X$ is a sheaf of ideals. If $A \subset X$ is an analytic set we define the *gap sheaf of I with respect to* A by

$$I[A](U) := \{f \in \mathcal{O}_X(U): f|(U\setminus A) \in I(U\setminus A)\}$$

for $U \subset X$ open. Obviously we have

$\mathcal{I} \subset \mathcal{I}[A] \subset \mathcal{O}_X$.

Proposition. If X is a complex space, $\mathcal{I} \subset \mathcal{O}_X$ is a coherent ideal, $A \subset X$ is an analytic set and $\mathcal{A} \subset \mathcal{O}_X$ is any coherent ideal such that $A = \operatorname{supp}(\mathcal{O}_X/\mathcal{A})$. Then

$$\mathcal{I}[A] = \bigcup_{k=0}^{\infty} \mathcal{I}:\mathcal{A}^k.$$

In particular, the gap sheaf $\mathcal{I}[A]$ is coherent.

Proof. Put

$$\mathcal{J} := \bigcup_{k=0}^{\infty} \mathcal{I}:\mathcal{A}^k \subset \mathcal{O}_X.$$

$\mathcal{A}$ is coherent by 0.13, hence $\mathcal{I}:\mathcal{A}^k$ is coherent for any k (0.41), and $\mathcal{J}$ is coherent by 0.40.

If $p \in X \setminus A$, then $\mathcal{A}_p^k = \mathcal{O}_{X,p}$ for any k; hence $\mathcal{I}_p : \mathcal{A}_p^k = \mathcal{I}_p$ and consequently $\mathcal{I}_p = \mathcal{J}_p$. This proves

$$\mathcal{J} \subset \mathcal{I}[A].$$

In order to prove the opposite inclusion, take $U \subset X$ open and $f \in \mathcal{I}[A](U)$. Then

$$\mathcal{B} := \mathcal{I}:\mathcal{O}_U f \subset \mathcal{O}_U$$

is a coherent sheaf of ideals. If $p \in U \setminus A$, then $f_p \in \mathcal{I}_p$ and $\mathcal{B}_p = \mathcal{O}_{X,p}$. Hence

$$\operatorname{supp}(\mathcal{O}_U/\mathcal{B}) \subset A \cap U.$$

The proposition from 0.15 implies $\mathcal{A}^k|V \subset \mathcal{B}|V$ for some $k \in \mathbb{N}$ and $V \subset U$. Consequently for any $x \in V$ we have

$$\mathcal{A}_x^k \cdot f_x \subset \mathcal{I}_x, \text{ hence } f_x \in \mathcal{I}_x:\mathcal{A}_x^k \text{ and } f_x \in \mathcal{J}_x.$$

This proves $\mathcal{I}[A] \subset \mathcal{J}$.

Remark. In a similar way the gap sheaf of any $\mathcal{O}_X$-modules $\mathcal{F}$ in an $\mathcal{O}_X$-module $\mathcal{G} \supset \mathcal{F}$ can be defined and the coherence proof is almost the same. For more details on gap sheaves we refer to [144] and [137].

0.43. In order to give a "geometric" interpretation of the gap sheaf of a sheaf of ideals (0.44), we first introduce another useful concept. Let X be a complex space, $A \subset X$ an analytic set and $\mathcal{O}[A]$ the gap sheaf of the zero sheaf $\mathcal{O}$ in $\mathcal{O}_X$ relative to A. We call A *analytically rare*, if

$\mathcal{O}[A] = \mathcal{O}$.

More explicitely, this means that for any open $U \subset X$ the restriction map

$$\mathcal{O}_X(U) \to \mathcal{O}_X(U \setminus A)$$

is injective. This is obviously a local property of A.

We will use the following algebraic

Lemma. If $\mathfrak{a}$ is an ideal in a noetherian ring R, then Ann $\mathfrak{a} = 0$ implies that $\mathfrak{a}$ contains a non-zero-divisor.

Proof. Assume $\mathfrak{a}$ contains only zero-divisors. Then

$$\mathfrak{a} \subset \bigcup_{\mathfrak{p} \in \text{Ass } R} \mathfrak{p},$$

where Ass R is the set of prime ideals of the form

$$\mathfrak{p} = \text{Ann}(a)$$

with $a \in R$. Since Ass R is finite, $\mathfrak{a}$ must be contained in one such prime ideal, i.e. there is an $a \in R$ with $a \neq 0$ and

$\mathfrak{a} \subset \text{Ann}(a)$, hence $a \in \text{Ann } \mathfrak{a}$.

For details we refer to [14], Chap. IV.

Proposition. Let X be a complex space and $A \subset X$ an analytic set with the full sheaf of ideals $\mathcal{A} \subset \mathcal{O}_X$. Then the following conditions are equivalent:

i) A is analytically rare.

ii) $\mathcal{A}_p$ contains a non zero divisor for every $p \in X$.

iii) If $U \subset X$ is open and $\mathcal{I} \subset \mathcal{O}_U$ is any coherent ideal with

$$\text{supp}(\mathcal{O}_U/\mathcal{I}) \subset A \cap U$$

then $\mathcal{I}_p$ contains a non zero divisor for every $p \in U$.

Proof. i) ⇒ ii). If

$$\mathcal{O}[A] = \bigcup_{k=0}^{\infty} \mathcal{O}:\mathcal{A}^k = \mathcal{O},$$

we obtain Ann $\mathcal{A}_p = 0$ for every $p \in X$ and the assertion follows from the lemma.

ii) ⇒ iii). By HILBERTs Nullstellensatz we obtain

$$\mathcal{A}_p^k \subset \mathcal{I}_p$$

for some $k \in \mathbb{N}$ depending on $p \in X$. If $a \in A_p$ is a non-zero-divisor, then $a^k \in A_p^k$ is a non-zero-divisor.

iii) $\Rightarrow$ i). Assume there is an open $U \subset X$ and

$$0 \neq f \in \mathrm{Ker}(\mathcal{O}(U) \to \mathcal{O}(U \setminus A)).$$

Then we define

$$\mathcal{I} := \mathcal{A}nn(\mathcal{O}_U f) \subset \mathcal{O}_U.$$

If $x \in U \setminus A$, then $f_x = 0$ and $\mathcal{I}_x = \mathcal{O}_{X,x}$. Hence

$$\mathrm{supp}(\mathcal{O}_U/\mathcal{I}) \subset A \cap U.$$

On the other hand, since $f \neq 0$, there is a $p \in U$ such that $f_p \neq 0$. Consequently $\mathcal{I}_p$ consists of zero-divisors.

Remark. It is easy to show that an analytic subset A of a reduced complex space X is rare if and only if there is no irreducible component $X' \subset X$ with $X' \subset A$. A similar result can be obtained for arbitrary complex spaces by using their *primary decomposition* (see [136],[91]).

0.44. We want to prove a generalization of the so called Lemma of RITT (see [120],[148],[59]). For this purpose we need the following

Remark. Let X be a complex space, $\mathcal{J} \subset \mathcal{I} \subset \mathcal{O}_X$ coherent ideals and $A \subset X$ an analytic subset. If $Y \hookrightarrow X$ denotes the closed subspace generated by $\mathcal{I}$, then

a) $\mathcal{I}[A] = \mathcal{I}[A \cap Y] \subset \mathcal{O}_X$,

b) $\mathcal{I}[A]/\mathcal{J} = (\mathcal{I}/\mathcal{J})[A] \subset \mathcal{O}_X/\mathcal{J}$,

c) $\mathcal{I}[A]/\mathcal{J} = (\mathcal{I}/\mathcal{J})[A \cap Y] \subset \mathcal{O}_X/\mathcal{J}$.

Proof. a) is clear since $\mathcal{I}_p = \mathcal{I}[A]_p = \mathcal{O}_{X,p}$ for $p \in X \setminus Y$.

b) Take $p \in X$ and

$$f_p = F_p + \mathcal{J}_p \in \mathcal{O}_{X,p}/\mathcal{J}_p.$$

If $U \subset X$ is an open neighbourhood of p and $F \in \mathcal{O}_X(U)$ represents F_p, then

$$\begin{aligned} f_p \in \mathcal{I}[A]_p/\mathcal{J}_p &\Leftrightarrow F_x \in \mathcal{I}_x \text{ for all } x \in U \setminus A \\ &\Leftrightarrow f_x \in \mathcal{I}_x/\mathcal{J}_x \text{ for all } x \in U \setminus A \\ &\Leftrightarrow f_p \in ((\mathcal{I}/\mathcal{J})[A])_p. \end{aligned}$$

c) follows from a) and b).

RITTs Lemma. Let X be a complex space, $Y \hookrightarrow X$ a closed complex subspace and $A \subset X$ an analytic subset. Denote by $\mathcal{I} \subset \mathcal{O}_X$ the sheaf of ideals defining Y. Then the closed subspace $Y' \hookrightarrow X$ defined by the gap sheaf $\mathcal{I}[A]$ has the following properties:

a) Y' is a closed subspace of Y.

b) $|Y'| = \overline{|Y| \setminus A}$

c) $Y' \cap (X \setminus A) = Y \cap (X \setminus A)$ and Y' is minimal with respect to this property, i.e. if $Z \hookrightarrow X$ is any closed subspace such that

$$Z \cap (X \setminus A) = Y \cap (X \setminus A),$$

then Y' is a closed subspace of Z.

d) $A \cap Y'$ is analytically rare in Y'.

e) $A \cap Y$ is analytically rare in Y if and only if $Y' = Y$.

We call Y' the *closure of* $Y \setminus A$ *in* X and denote it by

$$\mathrm{clos}_X(Y \setminus A).$$

Proof. a) is clear since $\mathcal{I} \subset \mathcal{I}[A]$.

c) follows immediately from the definition of the gap sheaf.

b) Put $B := \overline{|Y| \setminus A}$. Assertion c) implies $B \subset |Y'|$.

Conversely, if $p \in A \setminus B$, there is a neighbourhood U of p such that $U \cap (Y \setminus A) = \emptyset$. Hence $\mathcal{I}[A]_p = \mathcal{O}_{X,p}$ and $p \notin Y'$.

d) and e) are an easy consequence of the above remark. First we obtain

$$\mathcal{I}[A]/\mathcal{I} = \mathcal{O}[A \cap Y] \subset \mathcal{O}_X/\mathcal{I},$$

hence $Y' \hookrightarrow Y$ is the subspace generated by $\mathcal{O}[A \cap Y] \subset \mathcal{O}_Y$. This implies e). Furthermore we obtain

$$\mathcal{O} = \mathcal{O}[A \cap Y]/\mathcal{O}[A \cap Y] = \mathcal{O}[A \cap Y][A \cap Y]/\mathcal{O}[A \cap Y] = \mathcal{O}[A \cap Y'],$$

and d) is proved.

As an immediate consequence we note:

Identity theorem for complex subspaces. Let X be a complex space with closed subspaces Y and Z and let $A \subset X$ be an analytic set. If A is analytically rare in Y and if

$$Y \setminus A \hookrightarrow Z \setminus A$$

is a closed subspace, then

$$Y \hookrightarrow Z$$

is a closed subspace. In particular, if A is analytically rare in Y and Z, then

$$Y \setminus A = Z \setminus A \text{ implies } Y = Z.$$

0.45. Let X be a complex space with a closed subspace $Y \hookrightarrow X$, defined by the coherent ideal $I \subset \mathcal{O}_X$. Y is called a *hypersurface* (*in* X), if for every $p \in Y$ there is a non-zero-divisor $a \in \mathcal{O}_{X,p}$ such that

$$I_p = \mathcal{O}_{X,p} \cdot a.$$

Recall that an $\mathcal{O}_X$-module F is called *invertible*, if for every $p \in X$ there is an open neighbourhood U and an isomorphism

$$\mathcal{O}_X|U \to F|U.$$

Then obviously Y is a hypersurface if and only if I is invertible.

The so called "active Lemma" [50] implies

$$\dim_p Y = \dim_p X - 1$$

for every $p \in Y$ in case Y is a hypersurface. However, the converse is not true.

Example. Take $\mathbb{C}^2$ with coordinate functions x,y and let $X \hookrightarrow \mathbb{C}^2$ (resp. $Y \hookrightarrow \mathbb{C}^2$) be the closed subspace defined by $x \cdot y$, y^2 (resp. $x \cdot y$, y^2, x). Then

$$\dim_0 Y = 0, \ \dim_0 X = 1,$$

the ideal of $Y \hookrightarrow X$ is generated by the residue class of x, and Y is not a hypersurface in X (Fig. 1).

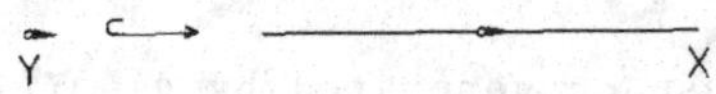

Figure 1

Recall that a sheaf of ideals $I \subset \mathcal{O}_X$ is called *principal*, if for every $p \in X$ there is an $a \in \mathcal{O}_{X,p}$ such that $I_p = \mathcal{O}_{X,p} \cdot a$.

Proposition. Let X be a complex space with a closed subspace $Y \hookrightarrow X$ defined by a principal ideal $I \subset \mathcal{O}_X$. Define

$$X' := \mathrm{clos}_X(X \setminus Y) \text{ and } Y' := Y \cap X'$$

(see 0.27). Then $Y' \hookrightarrow X'$ is a hypersurface.

More explicitely, $X' \hookrightarrow X$ is the closed subspace defined by the coherent ideal

$$\bigcup_{k=0}^{\infty} Ann\,(I^k) \subset \mathcal{O}_X.$$

Proof. For the ideal defining $X' \hookrightarrow X$ we have

$$\mathcal{O}[Y] = \bigcup \mathcal{O}:\mathcal{I}^k = \bigcup Ann\ \mathcal{I}^k$$

by 0.42. If $p \in X'$ take a generator $a \in \mathcal{O}_{X,p}$ of $\mathcal{I}_p$. Since

$$\mathcal{O}[Y]_p = \bigcup \mathrm{Ann}(a^k) \neq \mathcal{O}_{X,p},$$

the residue class of a in $\mathcal{O}_{X',p}$ is a non-zero-divisor. By definition of the analytic intersection of Y and X' it generates the ideal of Y' in X' at p and the assertion is proved.

0.46. Let X be a complex space. A subset $C \subset X$ is called *constructible* if one can find an $n \in \mathbb{N}$ and analytic subsets $A_1,\ldots,A_n$, $B_1,\ldots,B_n$ of X such that

$$C = \bigcup_{i=1}^{n} (A_i \setminus B_i) .$$

It is an easy exercise to show that the constructible sets form the smallest class of subsets of X satisfying the following conditions:

a) $C \subset X$ analytic $\Rightarrow$ C constructible.

b) $C_1,\ldots,C_k \subset X$ constructible $\Rightarrow$ $C_1 \cap \ldots \cap C_k$ and $C_1 \cup \ldots \cup C_k$ constructible.

c) $C \subset X$ constructible $\Rightarrow$ $X \setminus C$ constructible.

An often used method to prove the analyticity of a certain subset is to prove first it is constructible. Then it is sufficient to prove the set is closed:

Proposition. Every closed constructible subset C of a complex space X is analytic.

Proof. Assume

$$C = \bigcup_{i=1}^{n} (A_i \setminus B_i) .$$

If C is closed then

$$C = \bigcup_{i=1}^{n} \overline{(A_i \setminus B_i)} .$$

By the Lemma of RITT (0.44) every set $\overline{A_i \setminus B_i}$ is analytic in X. Hence C is analytic.

CHAPTER 1. COHERENT SHEAVES

The main object of this chapter is to state several coherence theorems for image sheaves and to prove some often used consequences. To begin with we derive a "geometric picture" of coherent sheaves.

1.1. Let S be a fixed complex space. We consider the *category of complex spaces over* S. Its *objects* are pairs

(X,φ), where $\varphi: X \to S$

is a holomorphic map. If (X,φ) and (X',φ') are two complex spaces over S, then

$\mathrm{Hol}_S(X,X')$

denotes the set of *morphisms* (or *holomorphic maps over* S, i.e. holomorphic maps $\chi: X \to X'$ such that the diagram

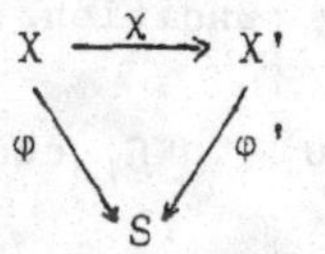

commutes). It is easy to see that such a map is *fibre preserving*, i.e. for any $s \in S$ there is a restriction

$\chi_s: X_s \to X'_s$

of χ (compare 0.17 and 0.27).

In the following section we will furnish complex spaces over S with several kinds of algebraic structures.

1.2. First we study cones over a complex space S. This will be useful for a proof of CHOWs Theorem in chapter 4.

For any n consider the scalar multiplication

$\mathbb{C} \times \mathbb{C}^n \to \mathbb{C}^n$, $(t,z_1,\ldots,z_n) \mapsto (tz_1,\ldots,tz_n)$.

Its product with the identity on S yields a holomorphic map

$\mu: \mathbb{C} \times S \times \mathbb{C}^n \to S \times \mathbb{C}^n$,

which we call *scalar multiplication over* S. A closed complex subspace $X \hookrightarrow S \times \mathbb{C}^n$ is called *cone over* S, if it is invariant under the scalar multiplication, i.e. if there is a commutative diagram

$$\begin{array}{ccc} \mathbb{C} \times S \times \mathbb{C}^n & \xrightarrow{\ \mu\ } & S \times \mathbb{C}^n \\ \uparrow & & \uparrow \\ \mathbb{C} \times X & \overset{\mu'}{\dashrightarrow} & X \end{array}$$

with a holomorphic map μ' (which is the restriction of μ).

If $X \hookrightarrow S \times \mathbb{C}^n$ is a cone then for any fixed $\lambda \in \mathbb{C}^*$ the automorphism

$$\mathbb{C}^n \to \mathbb{C}^n, \ (z_1,\ldots,z_n) \mapsto (\lambda z_1,\ldots,\lambda z_n)$$

induces automorphisms

$$\mu_\lambda : S \times \mathbb{C}^n \to S \times \mathbb{C}^n \text{ and } \mu'_\lambda : X \to X.$$

For the definition of an "abstract cone" X over S and a possible realization as a subspace of $S \times \mathbb{C}^n$ we refer to [113].

If $U \subset S$ is open we consider any polynomial of

$$\mathcal{O}_S(U)[z_1,\ldots,z_n]$$

as a holomorphic function on $U \times \mathbb{C}^n$. The following result is due to H. CARTAN (see [118]).

Proposition. Let S be any complex space and assume $X \hookrightarrow S \times \mathbb{C}^n$ is a closed complex subspace defined by the coherent ideal $\mathcal{I} \subset \mathcal{O}_{S\times\mathbb{C}^n}$. Then the following conditions are equivalent:

i) X is a cone over S.

ii) For any $s \in S$ one can find an open neighbourhood U in S and homogeneous polynomials

$$H_1,\ldots,H_r \in \mathcal{O}_S(U)[z_1,\ldots,z_n]$$

generating $\mathcal{I}$ on $U \times \mathbb{C}^n$.

Proof. Denote by $\mathcal{J} \subset \mathcal{O}_{\mathbb{C}\times S\times\mathbb{C}^n}$ the ideal defining the subspace $\mathbb{C} \times X \hookrightarrow \mathbb{C} \times S \times \mathbb{C}^n$. For any open $U \subset S$ the projection $\mathbb{C} \times U \to U$ determines a monomorphism

$$\mathcal{O}_S(U) \to \mathcal{O}_{\mathbb{C}\times S}(\mathbb{C} \times U).$$

Hence we may consider

$$\mathcal{O}_S(U)[z_1,\ldots,z_n] \subset \mathcal{O}_{\mathbb{C}\times S}(\mathbb{C} \times U)[z_1,\ldots,z_n]$$

as a subring. If

$$H_1,\ldots,H_r \in \mathcal{O}_S(U)[z_1,\ldots,z_n]$$

generate $\mathcal{I}$ on $U \times \mathbb{C}^n$, then the same polynomials, as elements of

$\mathcal{O}_{\mathbb{C}\times S}(\mathbb{C} \times U)[z_1,\ldots,z_n]$ generate J on $\mathbb{C} \times U \times \mathbb{C}^n$. Denote by

$$\mathcal{O}_S(U)[z_1,\ldots,z_n] \to \mathcal{O}_{\mathbb{C}\times S}(\mathbb{C} \times U)[z_1,\ldots,z_n],\ H \mapsto H^\bullet,$$

the homomorphism determined by the scalar multiplication over U. The relation between H and $H^\bullet$ may be expressed by

$$H^\bullet(t,s,z) = H(s,t\cdot z).$$

If $H \in \{H_1,\ldots,H_r\}$ is homogeneous of degree k we obtain

$$H^\bullet = t^k \cdot H.$$

By using 0.17 this implies ii) ⇒ i).
In order to prove the less obvious implication i) ⇒ ii), fix $s \in S$ and denote by

$$R := \mathcal{O}_{S,s}[\langle z_1,\ldots,z_n\rangle],\ R' := \mathcal{O}_{S,s}[\langle t,z_1,\ldots,z_n\rangle]$$

the local rings of $S \times \mathbb{C}^n$ resp. $\mathbb{C} \times S \times \mathbb{C}^n$ in (s,o) resp. (o,s,o). By

$$I \subset R \text{ and } J \subset R'$$

we denote the stalks of I and J, and by

$$\alpha: R \to R',\ h \mapsto h^\bullet,$$

we denote the homomorphism induced by the scalar multiplication. Our assumption i) implies

$$\alpha(I) \subset J.$$

If we consider $R \subset R'$ as a subring, any subset

$$\{h_1,\ldots,h_r\} \subset R$$

of generators of I is also a set of generators of J. As above, we may express the relation between $h \in R$ and $h^\bullet = \alpha(h) \in R'$ by

$$h^\bullet(t,s,z) = h(s,t\cdot z).$$

If $h \in \{h_1,\ldots,h_r\}$, then $h^\bullet \in J$ implies that there are $\varphi_1,\ldots,\varphi_r \in R'$ such that

$$h^\bullet = \varphi_1 h_1 + \ldots + \varphi_r h_r.$$

For $\rho = 1,\ldots,r$ there is an expansion

$$\varphi_\rho = \sum_{i=0}^{\infty} \varphi_\rho^{(i)} t^i \text{ with } \varphi_\rho^{(i)} \in R.$$

Hence

$$h^{\bullet} = \sum_{i=0}^{\infty} (\varphi_1^{(i)} h_1 + \ldots + \varphi_r^{(i)} h_r) t^i. \qquad (*)$$

On the other hand there is an expansion

$$h = \sum_{i=0}^{\infty} h^{(i)},$$

where $h^{(i)} \in \mathcal{O}_{S,s}[z_1,\ldots,z_n]$ is homogeneous of degree i. The definition of $h^{\bullet}$ yields

$$h^{\bullet} = \sum_{i=0}^{\infty} h^{(i)} t^i \qquad (**)$$

and equating the coefficients of (*) and (**) we obtain

$$h^{(i)} \in I \text{ for } i = 0,\ldots,\infty.$$

Denote by $I' \subset R$ the ideal generated by $h_\rho^{(i)}$ for $\rho = 1,\ldots,r$ and $i = 0,\ldots,\infty$. As we have seen, $I' \subset I$.

In order to prove $I \subset I'$ we fix $h \in \{h_1,\ldots,h_r\}$. Then we may choose open neighbourhoods U of s in S and Z of o in $\mathbb{C}^n$ and representatives

$$H \in \mathcal{O}_{S\times\mathbb{C}^n}(U \times Z) \text{ of } h,$$

$$H^{(i)} \in \mathcal{O}_S(U)[z_1,\ldots,z_n] \text{ of } h^{(i)}$$

such that

$$H = \sum_{i=0}^{\infty} H^{(i)}$$

converges compactly in $U \times Z$. Consider the canonical map

$$\mathcal{O}_{S\times\mathbb{C}^n}(U \times Z) \to \mathcal{O}_{S,s}[\langle z_1,\ldots,z_n\rangle] = R.$$

It becomes continuous if we provide $\mathcal{O}_{S\times\mathbb{C}^n}(U \times Z)$ with the Fréchet topology of compact convergence and R with the Krull topology. Since $I' \subset R$ is closed in the Krull topology and

$$h = \sum_{i=0}^{\infty} h^{(i)}$$

we conclude $h \in I'$. This finally implies $I = I'$.

R being noetherian there is a finite number of homogeneous polynomials of $\mathcal{O}_{S,s}[z_1,\ldots,z_n]$, again denoted by $h_1,\ldots,h_r$, generating I. We may choose an open neighbourhood U of s in S and representatives

$$H_1,\ldots,H_r \in \mathcal{O}_S(U)[z_1,\ldots,z_n].$$

If $I' \subset \mathcal{O}_{U\times\mathbb{C}^n}$ denotes the sheaf of ideals generated by $H_1,\ldots,H_r$ the coherence of I and I' implies that there is an open neighbourhood Z of o in $\mathbb{C}^n$ such that

$$I = I' \text{ on } U \times Z,$$

if we shrink U sufficiently. But the fact that X is a cone easily implies

$$I = I' \text{ on } U \times \mathbb{C}^n$$

(use the automorphisms μ_λ defined above).

1.3. Let S be a complex space and $X \to S \times \mathbb{C}^{n+1}$ a cone over S. We shall indicate how to associate to X a "projective variety over S". By

$$\mathbb{P}_n(\mathbb{C}) = \{(z_o: z_1:\ldots:z_n): (z_o,\ldots,z_n) \in \mathbb{C}^{n+1}\setminus\{0\}\}$$

we denote the n-dimensional complex projective space where

$$(z_o:z_1:\ldots:z_n) := \mathbb{C}\cdot(z_o,\ldots,z_n)$$

is a line through the origin. We define

$$V_\nu := \{(z_o:\ldots:z_n) \in \mathbb{P}_n(\mathbb{C}): z_\nu \neq 0\} \text{ for } \nu = 0,\ldots,n, \text{ and}$$

$$\varphi_\nu: \mathbb{C}^n \to V_\nu,\ (x_1,\ldots,x_n) \mapsto (x_1:\ldots:x_\nu:1:x_{\nu+1}:\ldots:x_n).$$

It is well known how to provide $\mathbb{P}_n(\mathbb{C})$ with the structure of a complex manifold such that the maps φ_ν become biholomorphic (see e.g. [44]). Denote by $I \subset \mathcal{O}_{S\times\mathbb{C}^{n+1}}$ the ideal defining our given cone X. For a fixed $s \in S$ we choose an open neighbourhood U and homogeneous generators

$$H_1,\ldots,H_r \in \mathcal{O}_S(U)[z_o,\ldots,z_n]$$

according to 1.2. Given $\rho \in \{1,\ldots,r\}$ and $\nu \in \{0,\ldots,n\}$ there is a uniquely determined polynomial

$$H_{\rho,\nu}(x_1,\ldots,x_n) \in \mathcal{O}_S(U)[x_1,\ldots,x_n] \text{ such that}$$

$$H_\rho = z_\nu^k\cdot H_{\rho,\nu}\left(\frac{z_o}{z_\nu},\ldots,\frac{z_{\nu-1}}{z_\nu}, \frac{z_{\nu+1}}{z_\nu},\ldots,\frac{z_n}{z_\nu}\right),$$

where k denotes the degree of H_ρ. Via the isomorphism φ_ν we may consider the polynomials $H_{1,\nu},\ldots,H_{r,\nu}$ as holomorphic functions on $U\times V_\nu$.

Denote by

$$I_\nu \subset \mathcal{O}_{U\times V_\nu}$$

the sheaf of ideals generated by these polynomials.
Our construction implies

$$I_\nu = I_\mu \text{ on } U \times (V_\nu \cap V_\mu).$$

Hence the sheaves I_ν match together to an ideal

$$J_U \subset \mathcal{O}_{U\times \mathbb{P}_n(\mathbb{C})},$$

and if we take a covering of S by such open sets U we finally obtain a coherent ideal

$$J \subset \mathcal{O}_{S\times \mathbb{P}_n(\mathbb{C})}$$

which is independent of all the choices made. The corresponding closed complex subspace

$$\mathbb{P}(X) \hookrightarrow S \times \mathbb{P}_n(\mathbb{C})$$

is called the *projective variety over* S *associated to* X.
In chapter 4 we shall see that any closed complex subspace of $S \times \mathbb{P}_n(\mathbb{C})$ can be obtained in this way (Theorem of CHOW).

<u>1.4.</u> Now we introduce stronger algebraic structures on complex spaces over a given complex space S. For the categorial background of these procedures we refer e.g. to [109]. If

$$\alpha_{\mathbb{C}}, \mu_{\mathbb{C}}: \mathbb{C} \times \mathbb{C} \to \mathbb{C}$$

denote addition and multiplication in $\mathbb{C}$, we define

$$R = S \times \mathbb{C},\ \alpha = id_S \times \alpha_{\mathbb{C}},\ \mu = id_S \times \mu_{\mathbb{C}};$$

R together with the compositions

$$\alpha,\mu: R \times_S R = S \times \mathbb{C} \times \mathbb{C} \to R$$

is a commutative ring with unit in the category of complex spaces over S. The "ring axioms" are expressed by the commutativity of the obvious diagrams. The *unit* is the section

$$(id_S, 1): S \to S \times \mathbb{C},$$

where $1: S \to \mathbb{C}$ denotes the constant holomorphic map with value 1.

By definition, a *linear space over a complex space* S (or simply a *linear fibre space*) is a unitary $S \times \mathbb{C}$-module in the category of complex spaces over S, i.e. a complex space $L \to S$ over S together with compositions

$$+: L \times_S L \to L \quad \text{and}$$

$$\cdot: (S\times\mathbb{C}) \times_S L = \mathbb{C} \times L \to L$$

such that the module axioms hold.

In particular there is a (uniquely determined) *zero section*

$$0: S \to L.$$

The axioms imply that every fibre L_s ($s \in S$) is a $\mathbb{C}$-module in the category of complex spaces. By a theorem of CARTIER [110] L_s is a reduced complex space and hence isomorphic to some $\mathbb{C}^n$, where n depends on s.

It should be noted, that the converse is not true. The reader will easily find an example of a complex space L over a double point S with compositions + and $\cdot$ which do not satisfy the module axioms but induce the structure of an honest $\mathbb{C}$-vector space on the fibre L_s for the single point $s \in S$ (just make something non-linear in the nilpotent part). For the same reason morphisms of linear fibre spaces have to be defined by diagrams.

Let $(L,+,\cdot)$ and $(L',+',\cdot')$ be linear spaces over S. A holomorphic map $\xi: L \to L'$ over S is called a *homomorphism (of linear spaces over S)*, if the diagrams

$$\begin{array}{ccc} L \times_S L & \xrightarrow{\xi \times_S \xi} & L' \times_S L' \\ \downarrow{+} & & \downarrow{+'} \\ L & \xrightarrow{\xi} & L' \end{array}$$

and

$$\begin{array}{ccc} \mathbb{C} \times L & \xrightarrow{id \times \xi} & \mathbb{C} \times L' \\ \downarrow{\cdot} & & \downarrow{\cdot'} \\ L & \xrightarrow{\xi} & L' \end{array}$$

commute.

Then for any $s \in S$ the restrictions

$$\xi_s: L_s \to L'_s$$

are homomorphisms of $\mathbb{C}$-vector spaces but in general this is not sufficient for ξ to be a homomorphism.

Finally a closed complex subspace $L' \hookrightarrow L$ is called a *linear subspace* if the compositions of L can be restricted to L', and L' becomes a linear space over S in this way.

<u>1.5.</u> If L and L' are linear spaces over a complex space S, we denote by

$$\mathrm{Hom}_S(L,L')$$

the *set of homomorphisms* from L to L'.

By 0.19 we may consider every $f \in \mathcal{O}_S(S)$ as a holomorphic map $S \to \mathbb{C}$. If $\xi \in \mathrm{Hom}_S(L,L')$, we have a holomorphic map

$$((\mathrm{id}_S,f)\circ\pi,\xi): L \to (S\times\mathbb{C}) \times_S L',$$

where $\pi: L \to S$ denotes the projection. If we compose the above map with the multiplication in L', we obtain a homomorphism $f\cdot\xi \in \mathrm{Hom}_S(L,L')$, and in this way $\mathrm{Hom}_S(L,L')$ becomes an $\mathcal{O}_S(S)$-module.

If $U \subset S$ is open we denote by L_U and L'_U the inverse images of L and L' over U and we define the sheaf $\mathcal{H}om_S(L,L')$ of $\mathcal{O}_S$-modules by

$$\mathcal{H}om_S(L,L')(U) = \mathrm{Hom}_U(L_U,L'_U).$$

In particular, we call

$$\mathcal{L}_S(L) := \mathcal{H}om_S(L,S\times\mathbb{C})$$

the *sheaf of linear forms on* L. Clearly $\mathcal{L}_S$ becomes a contravariant functor from the category of linear spaces over S into the category of sheaves of $\mathcal{O}_S$-modules. We write $\mathcal{L}$ instead of $\mathcal{L}_S$, if it is clear what S is.

<u>1.6.</u> <u>Duality Theorem.</u> Let S be a complex space. Then for any linear space L over S the sheaf $\mathcal{L}(L)$ of linear forms is a coherent $\mathcal{O}_S$-module and the functor $\mathcal{L}$ from the category of linear spaces over S to the category of coherent $\mathcal{O}_S$-modules is an antiequivalence.

For the <u>proof</u> we use

<u>Lemma 1.</u> Let L be a linear space over S. Then for any point $s \in S$ there is an open neighbourhood U of s in S such that L_U is isomorphic to a linear subspace of $U \times \mathbb{C}^n$ for some n.

For the proof of Lemma 1 we refer to [113].

We clearly have a canonical isomorphism

$$\mathrm{Hom}_S(S\times\mathbb{C}^n\ ,\ S\times\mathbb{C}^m) \to M(m\times n;\ \mathcal{O}_S(S)),$$

i.e. every homomorphism $\xi\colon S\times\mathbb{C}^n \to S\times\mathbb{C}^m$ is given by a holomorphic $(m\times n)$-matrix (ξ_{ij}) on S. We define the *kernel* of ξ, denoted by Ker ξ, as the complex subspace of $S\times\mathbb{C}^n$ generated by the holomorphic functions

$$\xi_{i1}z_1 + \ldots + \xi_{in}z_n \in \mathcal{O}(S\times\mathbb{C}^n)$$

where $i = 1,\ldots,m$ and $z_1,\ldots,z_n$ denote the coordinate functions in $\mathbb{C}^n$. Since they are linear in $z_1,\ldots,z_n$, Ker $\xi \hookrightarrow X\times\mathbb{C}^n$ is a linear subspace.

Lemma 2. Let $L \hookrightarrow S\times\mathbb{C}^n$ be a linear subspace. Then for any $s\in S$ there is an open neighbourhood U of s in S and a homomorphism $\xi\colon U\times\mathbb{C}^n \to U\times\mathbb{C}^m$ such that $L_U = \mathrm{Ker}\,\xi$.

The proof of Lemma 2 uses Proposition 1.2 (see [31]).

Now the coherence of $\mathcal{L}(L)$ is clear.
If $s\in S$, the lemmas give us an open neighbourhood U of s in S and a sequence

$$0 \to L_U \to U\times\mathbb{C}^n \to U\times\mathbb{C}^m$$

of homomorphisms of linear fibre spaces. If we apply the functor $\mathcal{L}$, we obtain

$$0 \leftarrow \mathcal{L}(L_U) \leftarrow (\mathcal{O}_S|U)^n \leftarrow (\mathcal{O}_S|U)^m$$

and an elementary computation shows, that this sequence is exact.

The proof that $\mathcal{L}$ is an antiequivalence is straightforward ([52], [43], [31]). We only mention the construction of a linear fibre space $\underline{V}(F)$ for a coherent $\mathcal{O}_S$-module F. We assume that there is an exact sequence

$$\mathcal{O}_S^m \xrightarrow{\ \eta\ } \mathcal{O}_S^n \to F \to 0.$$

Then η is given by a holomorphic $(n\times m)$-matrix on S. Its transpose defines a homomorphism

$$\xi\colon S\times\mathbb{C}^n \to S\times\mathbb{C}^m \text{ and we define } \underline{V}(F) := \mathrm{Ker}\,\xi.$$

We leave it as an exercise to show that this definition does not depend on the choice of the sequence and to glue together the local pieces.

If F,F' are coherent $\mathcal{O}_S$-modules and L,L' are linear spaces over S, the

Duality Theorem yields canonical isomorphisms

$$F \to L(\underline{V}(F)),\ L \to \underline{V}(L(L))$$

$$Hom_{O_S}(F,F') \to Hom_S(\underline{V}(F'),\underline{V}(F))$$

$$Hom_S(L,L') \to Hom_{O_S}(L(L'),L(L)).$$

<u>Corollary.</u> If L is a linear space over a complex space S, then the sheaf

$$S(L) := Hom_S(S\times C,L)$$

of *holomorphic sections* in L is a coherent O_S-module.

In case L is a vector bundle, the sheaf $S(L)$ is locally free. We recall that the covariant functor S from the category of vector bundles over S to the category of locally free O_S-modules is an equivalence [57]. However it does not extend to linear fibre spaces and coherent sheaves. It seems to be unknown if there is a canonical functor red in the category of linear fibre spaces.

<u>1.7.</u> We shall now define a *change of base* for linear fibre spaces. If $\varphi: S' \to S$ is a holomorphic map and L is a linear space over S, then we define

$$\varphi^*L := L \times_S S'.$$

The compositions in L induce compositions in φ^*L and φ^*L becomes a linear space over S'. Clearly φ^* is a covariant functor from linear spaces over S to linear spaces over S'. We have also denoted by φ^* the functor *analytic inverse image* (0.10). This is justified by the

<u>Proposition.</u> If $\varphi: S' \to S$ is a holomorphic map and L is a linear space over S, then there is a canonical isomorphism

$$\varphi^* L_S(L) \to L_{S'}(\varphi^*L). \qquad (*)$$

<u>Proof.</u> The pull back of linear forms yields a canonical homomorphism

$$L_S(L) \to \varphi_* L_{S'}(\varphi^*L)$$

and we get our homomorphism (*) by 0.10. To prove that it is an isomorphism we may assume that L is the kernel of a homomorphism

$$\xi: S \times \mathbb{C}^n \to S \times \mathbb{C}^m.$$

We apply the functor φ^* and conclude that φ^*L is the kernel of

$$\varphi^*\xi: S' \times \mathbb{C}^n \to S' \times \mathbb{C}^m.$$

Applying φ^* to the exact sequence

$$\mathcal{O}_S^m \to \mathcal{O}_S^n \to \mathcal{L}_S(L) \to 0$$

yields the exact sequence

$$\mathcal{O}_{S'}^m \to \mathcal{O}_{S'}^n \to \varphi^*\mathcal{L}_S(L) \to 0.$$

But since φ^*L was the kernel of $\varphi^*\xi$, this cokernel is the same as $\mathcal{L}_{S'}(\varphi^*L)$.

Since φ^* was right exact for sheaves, we get the

Corollary. The change of base φ^* of linear fibre spaces is a left exact functor.

1.8. We apply the change of base in the particular case where S' is a point $s \in S$. Let L be a linear fibre space over S and put $\mathcal{F} := \mathcal{L}_S(L)$. Now

$$\varphi^*\mathcal{F} = \mathcal{F}_s \otimes_{\mathcal{O}_{S,s}} \mathbb{C} = \mathcal{F}_s/m_{S,s}\mathcal{F}_s,$$

and

$$\mathcal{L}_{S'}(\varphi^*L) = \mathrm{Hom}_{\mathbb{C}}(L_s,\mathbb{C}) =: (L_s)^*.$$

Hence we have a canonical isomorphism

$$\mathcal{F}_s/m_{S,s}\mathcal{F}_s \to (L_s)^*.$$

Conversely, if $\mathcal{F}$ is a coherent $\mathcal{O}_S$-module, and $L := \underline{V}_S(\mathcal{F})$ we obtain a canonical isomorphism

$$L_s \to (\mathcal{F}_s/m_{S,s}\mathcal{F}_s)^*.$$

Proposition. Let L be a linear fibre space over a reduced complex space S. If $s \to \dim_{\mathbb{C}} L_s$ is a locally constant function on S, then L is a holomorphic vector bundle.

Since L is a vector bundle if and only if $\mathcal{L}_S(L)$ is locally free, this follows from the

Lemma. Let S be a reduced complex space and $\mathcal{F}$ an $\mathcal{O}_S$-module of finite type. If

$$s \to \dim_{\mathbb{C}}(\mathcal{F}_s/m_{S,s}\mathcal{F}_s)$$

is a locally constant function on S, then $\mathcal{F}$ is locally free.

Proof. Take $s \in S$ and an open neighbourhood U of s with an exact sequence

$$0 \to K \to (\mathcal{O}_S|U)^n \to F|U \to 0$$

where $n = \dim_{\mathbb{C}}(F_x/m_{S,x}F_x)$ for all $x \in U$. If $V \subset U$ is open and

$$f = (f_1,\ldots,f_n) \in K(V)$$

we get $f_1(x) = \ldots = f_n(x) = 0$ for all $x \in V$, since n is the minimal number of generators of F_x. Since S is reduced, this implies $f = 0$ by 0.16, and hence $K = 0$.

1.9. Given a complex space S and a coherent $\mathcal{O}_S$-module F, we are going to associate to F a projective variety over S, denoted by $\mathbb{P}(F)$. Consider the linear space over S

$$\pi: \underline{V}(F) \to S,$$

associated to F. If $U \subset S$ is open such that there is an exact sequence

$$\mathcal{O}_U^m \to \mathcal{O}_U^n \to F|U \to 0,$$

then there is a dual exact sequence

$$U\times\mathbb{C}^m \overset{\xi}{\leftarrow} U\times\mathbb{C}^n \leftarrow V(F)|U \leftarrow 0,$$

where ξ is given by a matrix

$$(\xi_{ij}) \in M(m\times n;\ \mathcal{O}_S(U)).$$

In particular, $\underline{V}(F)|U \hookrightarrow U\times\mathbb{C}^n$ is a cone over U (see 1.2) and as in 1.3 we may define a projective variety over U

$$\mathbb{P}(\underline{V}(F)|U) \hookrightarrow U\times\mathbb{P}_{n-1}(\mathbb{C}).$$

This is the subspace determined by the homogeneous system of m linear equations with coefficients ξ_{ij}. Given a covering of S with open sets U as above it is obvious how to glue together the local pieces $\mathbb{P}(\underline{V}(F)|U)$ (see 0.24) and we obtain

$$\mathbb{P}(F) := \mathbb{P}(\underline{V}(F))$$

together with a canonical projection map

$$\mathbb{P}(\pi): \mathbb{P}(F) \to S.$$

It is called the *projective variety over* S *associated to* F.
For any $s \in S$ the fibre $\mathbb{P}(F)_s$ is a projective space and according to 1.8 we have

$$\dim \mathbb{P}(F)_s = \dim_{\mathbb{C}}(F_s/m_{S,s}F_s) - 1.$$

1.10. We recall some well known definitions from topology. Let $\varphi: X \to Y$ be a continuous map between locally compact topological spaces. φ is called

proper, if for any compact $K \subset Y$ the set $\varphi^{-1}(K) \subset X$ is also compact

finite, if it is proper and every point $p \in X$ is an isolated point in the fibre $\varphi^{-1}(\varphi(p))$, or equivalently, if it is closed and has finite fibres.

For the basic properties of proper maps we refer to [12]. The following elementary fact is often useful.

Remark. Let $\varphi: X \to Y$ be a continuous map between locally compact spaces and assume there is a point $q \in Y$ such that $\varphi^{-1}(q) \subset X$ is compact. Then there are open neighbourhoods U of $\varphi^{-1}(q)$ in X and V of q in Y such that

$$\varphi|U: U \to V$$

is proper.

The proof is an easy exercise. A little bit more delicate is the following

Lemma 1. Let $\varphi: X \to Y$ be a holomorphic map between complex spaces, $q \in Y$ and $A \subset X$ a connected component of $\varphi^{-1}(q)$. If A is compact, then there are open neighbourhoods U of A in X and V of q in Y such that

$$\varphi|U: U \to V$$

is proper [138].

Proof. Put $B := \varphi^{-1}(q) \setminus A$. Since $\varphi^{-1}(q)$ is an analytic set, A and B are closed in X (and consequently analytic). This is a consequence of the fact that locally $\varphi^{-1}(q)$ has only a finite number of connected components.

Choose a relatively compact open neighbourhood W of A such that

$$\partial W \cap A = \partial W \cap B = \emptyset.$$

Since $q \notin \varphi(\partial W)$ we may choose for V any open neighbourhood of q such that

$$V \cap \varphi(\partial W) = \emptyset \text{ and } U := \varphi^{-1}(V) \cap W = \varphi^{-1}(V) \cap \overline{W}$$

is as desired.

It should be noted that this lemma remains true for continuous maps between locally compact spaces, but then the proof is more complicated.

Somewhat similar is the proof of

Lemma 2. Let $\varphi: X \to Y$ be a proper holomorphic map and fix $q \in Y$. Then:

a) $\varphi^{-1}(q)$ has only a finite number of connected components.

b) If

$$\varphi^{-1}(q) = A_1 \cup \ldots \cup A_n$$

is the decomposition into connected components, then there are arbitrarily small open neighbourhoods V of q, such that

$$\varphi^{-1}(V) = U_1 \cup \ldots \cup U_n ,$$

where U_i is an open neighbourhood of A_i for $i = 1,..,n$ and this union is disjoint.

Proof. a) is clear since $\varphi^{-1}(q)$ is a compact analytic set and analytic sets are locally connected.
In order to prove b) choose compact and disjoint neighbourhoods $K_1,\ldots,K_n$ of $A_1,..,A_n$. If $(V_j)_{j\in\mathbb{N}}$ is a fundamental system of open neighbourhoods of q, then $\varphi^{-1}(\overline{V_j}) \subset K_1 \cup \ldots \cup K_n$ for j sufficiently large and $V = V_j$ is as desired.

Assertion b) immediately implies:

Lemma 3. Let $\varphi: X \to Y$ be a finite holomorphic map, F an $\mathcal{O}_X$-module and $q \in Y$. Then

$$(\varphi_* F)_q = \prod_{p \in \varphi^{-1}(q)} F_p .$$

1.11. Given a holomorphic map $\varphi: X \to Y$, fix a point $p \in X$ and put $q := \varphi(p)$. Via the canonical homomorphism

$$\tilde{\varphi}_p : \mathcal{O}_{Y,q} \to \mathcal{O}_{X,p}$$

(see 0.3) $\mathcal{O}_{X,p}$ is an $\mathcal{O}_{Y,q}$-module. $\mathcal{O}_{X,p}$ is called

finite over $\mathcal{O}_{Y,q}$, if it is a finitely generated $\mathcal{O}_{Y,q}$-module

quasi-finite over $\mathcal{O}_{Y,q}$, if $\dim_{\mathbb{C}} \mathcal{O}_{X,p}/m_{Y,q}\mathcal{O}_{X,p} < \infty$.

We recall a fundamental result from local analytic geometry (see [73]).

Theorem. If $\varphi: X \to Y$ is a holomorphic map, $p \in X$ and $q := \varphi(p)$, then the following conditions are equivalent:

i) $\mathcal{O}_{X,p}$ is finite over $\mathcal{O}_{Y,q}$.

ii) $\mathcal{O}_{X,p}$ is quasi-finite over $\mathcal{O}_{Y,q}$.

iii) p is an isolated point of its fibre X_q.

This is a consequence of the WEIERSTRASS preparation theorem and of HILBERTs Nullstellensatz.

In 3.2 we shall prove that condition iii) implies that φ is finite in a neighbourhood of p.

1.12. We are going to state the so called "Finite Coherence Theorem" which was first proved by GRAUERT and REMMERT [49, Satz 27]. For alternative proofs we refer to [73] and [107].

Theorem. Let S be a complex space. If $\varphi: X \to S$ is a finite holomorphic map and F is a coherent $\mathcal{O}_X$-module, then $\varphi_* F$ is a coherent $\mathcal{O}_S$-module. The functor

$$F \mapsto \varphi_* F$$

from the category of coherent $\mathcal{O}_X$-modules to the category of coherent $\mathcal{O}_S$-modules is exact, i.e. for any exact sequence

$$0 \to F' \to F \to F'' \to 0$$

of $\mathcal{O}_X$-modules the sequence

$$0 \to \varphi_* F' \to \varphi_* F \to \varphi_* F'' \to 0$$

is also exact.

It should be noted, that the functor φ^* need not be left exact if $\varphi: X \to S$ is finite (see 0.10, Example 2 and 3.13).

1.13. Corollary. Let $\varphi: X \to S$ be a finite holomorphic map and assume S is Stein. Then X is Stein.

The converse will be proved in 2.32.

Proof [8]. Let

$$0 \to F' \to F \to F'' \to 0$$

be an exact sequence of coherent $\mathcal{O}_X$-modules. By 1.11 the sequence

$$0 \to \varphi_* F' \to \varphi_* F \to \varphi_* F'' \to 0$$

of coherent $\mathcal{O}_S$-modules is exact. Since S is Stein

$$0 \to \varphi_* F'(S) \to \varphi_* F(S) \to \varphi_* F''(S) \to 0$$

is exact by Corollary 2 of 0.37. But

$$\varphi_* G(S) = G(X)$$

for any sheaf G on X. Hence we obtain the exact sequence

$$0 \to F'(X) \to F(X) \to F''(X) \to 0$$

and by using again Corollary 2 of 0.37 we conclude that X is Stein.

1.14. If $\varphi: X \to S$ is a finite holomorphic map then $\varphi_* \mathcal{O}_X$ is an $\mathcal{O}_S$-algebra which is coherent as $\mathcal{O}_S$-module. We are going to show that X can be reconstructed from $\varphi_* \mathcal{O}_X$. First we have to introduce some notations.

If S is a complex space , and A, A' are $\mathcal{O}_S$-algebras, we denote by

$$\mathrm{Hom}_{\mathcal{O}_S\text{-alg}}(A, A')$$

the set of homomorphisms $A \to A'$ of $\mathcal{O}_S$-algebras. If $\varphi: X \to S$ is a holomorphic map, A is an $\mathcal{O}_S$-algebra and B is an $\mathcal{O}_X$-algebra, then there is a canonical bijection [54]

$$\mathrm{Hom}_{\mathcal{O}_X\text{-alg}}(\varphi^* A, B) \to \mathrm{Hom}_{\mathcal{O}_S\text{-alg}}(A, \varphi_* B).$$

If $\psi: Z \to S$ is a second complex space over S, then there is a canonical map

$$\kappa: \mathrm{Hol}_S(X,Z) \to \mathrm{Hom}_{\mathcal{O}_S\text{-alg}}(\psi_* \mathcal{O}_Z, \varphi_* \mathcal{O}_X).$$

Let an arbitrary sheaf A of $\mathcal{O}_S$-algebras be given. Then a complex space $\zeta: Z \to S$ over S together with an isomorphism
$\alpha: A \to \zeta_* \mathcal{O}_Z$

of $\mathcal{O}_X$-algebras is called an *analytic spectrum* of A over S, if the map

$$\mathrm{Hol}_S(X,Z) \to \mathrm{Hom}_{\mathcal{O}_S\text{-alg}}(A, \varphi_* \mathcal{O}_X),\ \chi \mapsto \kappa(\chi) \circ \alpha$$

is bijective for every complex space $\varphi: X \to S$ over S. We write

$$Z = \mathrm{Specan}\ A$$

for short.

In case $S = \{s\}$ is a simple point, $A = A$ is a $\mathbb{C}$-algebra and if we put $X = \{s\}$, we obtain a bijective map

$$\mathrm{Hol}(\{s\}, \mathrm{Specan}\ A) \to \mathrm{Hom}_{\mathbb{C}\text{-alg}}(A, \mathbb{C}) = \mathrm{Spec}\ A.$$

In other words, in this case the underlying set of Specan A is the spectrum of the algebra A.

A sheaf A of $\mathcal{O}_S$-algebras is called *finite*, if A is a coherent $\mathcal{O}_S$-module. It should be noted that the condition that α be an isomorphism in the definition of the spectrum is too strong in case A is not finite but only of finite presentation. For this more general case we refer to [73].

<u>1.15. Theorem.</u> Let S be an arbitrary complex space.

a) For any finite $\mathcal{O}_S$-algebra A an analytic spectrum

$$\xi: \text{Specan } A \longrightarrow S$$

exists. Moreover the finiteness of A implies that ξ is a finite map and for every $s \in S$ there is a bijective map

$$n \longmapsto q$$

from the set of maximal ideals n of A_s to the set of points q of the fibre $(\text{Specan } A)_s$ together with a set of isomorphisms

$$(A_s)_n \longrightarrow \mathcal{O}_{(\text{Specan} A),q} \ .$$

b) The contravariant functor

$$X \longmapsto \varphi_* \mathcal{O}_X$$

from the category of finite complex spaces over S to the category of finite $\mathcal{O}_S$-algebras is a duality and the functor

$$A \longmapsto \text{Specan } A$$

is its quasi-inverse.

c) If A is a finite $\mathcal{O}_S$-algebra and $\varphi: T \to S$ is an arbitrary holomorphic map, then

$$\text{Specan } \varphi^* A = (\text{Specan } A) \times_S T \ .$$

The <u>proof</u> of this theorem may be found in [73], Exp. 19. We only sketch the local construction of the analytic spectrum.

Take $s \in S$ and $a_1, .., a_n \in A_s$ which generate A_s as an $\mathcal{O}_{S,s}$-algebra, hence we have a canonical epimorphism

$$\rho: \mathcal{O}_{S,s}[t_1, .., t_n] \to A_s \text{ with } \rho(t_i) = a_i \ .$$

For $i \in \{1, .., n\}$ fixed and $k \in \mathbb{N}$ consider the submodule

$$M_k = \mathcal{O}_{S,s} + \mathcal{O}_{S,s} \cdot a_i + \ldots + \mathcal{O}_{S,s} \cdot a_i^k \subset A_s \ .$$

Since A_s is a finitely generated $\mathcal{O}_{S,s}$-module, it is a noetherian module and hence

$$M_k = M_{k+1}$$

for k sufficiently large. Consequently there are $\lambda_o,\ldots,\lambda_k \in \mathcal{O}_{S,s}$ such that

$$a_i^{k+1} = \lambda_o + \lambda_1 a_i + \ldots + \lambda_k a_i^k .$$

This implies that the kernel of ρ contains monic polynomials

$$f_1 \in \mathcal{O}_{S,s}[t_1],\ldots,f_n \in \mathcal{O}_{S,s}[t_n].$$

Now ρ induces an epimorphism

$$\sigma: \mathcal{O}_{S,s}[t_1,\ldots,t_n]/(f_1,\ldots,f_n) \to A_s$$

of finitely generated $\mathcal{O}_{S,s}$-modules. Such an epimorphism may be extended to an open neighbourhood U of s, i.e. for i = 1,...,n we may find representatives

$$A_i \in A(U) \text{ and } F_i \in \mathcal{O}_S(U)[t_i]$$

of a_i and f_i such that $F_i(A_i) = 0$ and $F_1,\ldots,F_n$ are monic. Denote by

$$(F_1,\ldots,F_n) \subset \mathcal{O}_U[t_1,\ldots,t_n]$$

the sheaf of ideals generated by these sections. We leave it as an exercise to show that $\mathcal{O}_U[t_1,\ldots,t_n]/(F_1,\ldots,F_n)$ is a coherent (and even free) $\mathcal{O}_U$-module by using as generators the residue classes of the monomials

$$t_1^{r_1}\cdot\ldots\cdot t_n^{r_n} \text{ with } r_i < \deg F_i \text{ for } i = 1,\ldots,n.$$

By shrinking U we may further find an epimorphism

$$\mathcal{O}_U[t_1,\ldots,t_n]/(F_1,\ldots,F_n) \to A|U$$

inducing σ and $G_1,\ldots,G_m \in \mathcal{O}_S(U)[t_1,\ldots,t_n]$ such that the induced homomorphism

$$\mathcal{O}_U[t_1,\ldots,t_n]/(F_1,\ldots,F_n,G_1,\ldots,G_m) \to A|U$$

is an isomorphism.
By using the gadgets developed above we may give the local definition of Specan A. We assume S = U and define (for i = 1,...,n)

$$\mathcal{I}_i \subset \mathcal{O}_{S\times\mathbb{C}} \text{ resp. } \mathcal{I} \subset \mathcal{O}_{S\times\mathbb{C}^n} \text{ resp. } \mathcal{J} \subset \mathcal{O}_{S\times\mathbb{C}^n}$$

as the sheaf of ideals generated by the polynomials

$$F_i \text{ resp. } (F_1,\ldots,F_n) \text{ resp. } (F_1,\ldots,F_n,G_1,\ldots,G_m).$$

They determine closed complex subspaces

$Y_i \to S\times\mathbb{C}$ resp. $Y \to S\times\mathbb{C}^n$ resp. $Z \to S\times\mathbb{C}^n$

and as the restrictions of the canonical projections we obtain holomorphic maps

$\psi_i\colon Y_i \to S$ resp. $\psi\colon Y \to S$ resp. $\zeta\colon Z \to S$.

Since $F_1,\ldots,F_n$ are monic, $\psi_1,\ldots,\psi_n$ have finite fibres and the "continuity of roots" (see e.g. [73]) implies that $\psi_1,\ldots,\psi_n$ are closed, hence finite. Since

$Y = Y_1 \times_S \cdots \times_S Y_n$,

ψ is finite and since $Z \hookrightarrow Y$ is a closed subspace, ζ is finite. Now we define

Specan $A := Z$.

In order to construct an isomorphism

$\alpha\colon A \to \zeta_*\mathcal{O}_Z$

we consider the exact sequence over $S\times\mathbb{C}^n$

$$0 \to \mathcal{J} \to \mathcal{O}_{S\times\mathbb{C}^n} \to \mathcal{O}_Z' \to 0,$$

where $\mathcal{O}_Z'$ denotes the trivial extension of $\mathcal{O}_Z$ to $S\times\mathbb{C}^n$. Applying π_* to this sequence yields the second line of the following diagram of $\mathcal{O}_S$-modules:

$$\begin{array}{ccccccccc} 0 & \to & (F_1,\ldots,F_n,G_1,\ldots,G_m) & \hookrightarrow & \mathcal{O}_S[t_1,\ldots,t_n] & \to & A & \to & 0 \\ & & \downarrow & & \downarrow{\scriptstyle\beta} & & \downarrow{\scriptstyle\alpha} & & \\ 0 & \longrightarrow & \pi_*\mathcal{J} & \hookrightarrow & \pi_*\mathcal{O}_{S\times\mathbb{C}^n} & \longrightarrow & \zeta_*\mathcal{O}_Z = \pi_*\mathcal{O}_Z, & & \end{array}$$

where β denotes the canonical monomorphism. Since $\beta(F_1,\ldots,F_n,G_1,\ldots,G_m) \subset \pi_*\mathcal{J}$ according to the definition of $\mathcal{J}$, we obtain α induced by β, and it turns out that α is an isomorphism.

For the proof of b) we refer to [73]. The proof of c) is left to the reader (use the properties of the analytic spectrum to show the universal property of Specan φ^*A).

1.16. In general the functor φ_* is not right exact see (0.10 as) this is the case for a finite map φ. This defect is controlled by the "higher image sheaves". For details we refer to [49].

If $\varphi: X \to Y$ is a holomorphic map and F is a sheaf of $\mathcal{O}_X$-modules, we consider for $k \in \mathbb{N}$ the presheaf

$$V \to H^k(\varphi^{-1}(V), F),\ V \subset Y \text{ open},$$

of $\mathcal{O}_Y$-modules. For $k \geq 1$ this need not be a sheaf (why?) and we denote by

$$R^k\varphi_* F$$

the associated sheaf. Clearly $R^k\varphi_*$ is a covariant functor. The exactness of the cohomology sequence and the exactness of the inductive limit yield the

Proposition. For any exact sequence

$$0 \to F' \to F \to F'' \to 0$$

of $\mathcal{O}_X$-modules the induced sequence

$$0 \to \varphi_* F' \to \varphi_* F \to \varphi_* F'' \to R^1\varphi_* F' \to R^1\varphi_* F \to R^1\varphi_* F'' \to R^2\varphi_* F' \to \ldots$$

of $\mathcal{O}_Y$-modules is exact.

Lemma. If $\varphi: X \to Y$ is a finite holomorphic map and F is a coherent $\mathcal{O}_X$-module, then

$$R^k\varphi_* F = 0 \quad \text{for } k \geq 1 \qquad (*)$$

and there is a canonical isomorphism

$$H^k(Y, \varphi_* F) \to H^k(X, F) \quad \text{for } k \geq 0. \qquad (**)$$

Proof. If $S \subset Y$ is an open Stein subset then by 1.13 we have

$$H^k(\varphi^{-1}(S), F|\varphi^{-1}(S)) = 0 \quad \text{for } k \geq 1.$$

This implies (*) and assertion (**) follows by using the theorem of Leray.

It should be noted that for a not necessarily finite holomorphic map (**) is a consequence of (*) (see [49], Satz 6).

This is an easy consequence of 1.15 and Theorem B (0.37).

1.17. Proper coherence theorem. Let $\varphi: X \to Y$ be a proper holomorphic map and assume F is a coherent $\mathcal{O}_X$-module. Then for all $k \in \mathbb{N}$ the $\mathcal{O}_Y$-modules $R^k\varphi_*F$ are coherent.

In addition, if $S \subset Y$ is any open Stein subspace, then the canonical map

$$H^k(\varphi^{-1}(S), F) \to (R^k\varphi_*F)(S)$$

is an isomorphism.

The original proof is due to GRAUERT [42]. For simplified proofs we refer to [34, 86]. The bijectivity of the canonical map follows by using Theorem B and spectral sequence arguments [49].
For the special case where Y is a point we obtain the

Finiteness Theorem of Cartan-Serre. If X is a compact complex space and F is a coherent $\mathcal{O}_X$-module, then for $k \in \mathbb{N}$

$$\dim_{\mathbb{C}} H^k(X, F) < \infty.$$

Given a non-proper holomorphic map $\varphi: X \to Y$ the reader will easily find a coherent $\mathcal{O}_X$-module F such that φ_*F is not coherent

1.18. In its original form [117] REMMERTs mapping theorem states that the images of analytic sets under proper holomorphic maps are analytic. By using the proper coherence theorem one can even provide the image of a complex space with a distinguished complex structure.

REMMERTs mapping theorem. Let $\varphi: X \to Y$ be a proper holomorphic map and denote by $K \subset \mathcal{O}_Y$ the coherent sheaf of ideals which is the kernel of the canonical homomorphism

$$\tilde{\varphi}: \mathcal{O}_Y \to \varphi_*\mathcal{O}_X.$$

Then the complex subspace $\varphi(X) \hookrightarrow Y$ determined by K has the following properties:

a) $|\varphi(X)| = |\varphi|(X)$

b) There is a factorization

$$X \xrightarrow{\varphi'} \varphi(X) \hookrightarrow Y$$

of φ having the following universal property:
If $Y_1 \hookrightarrow Y$ is any closed complex subspace admitting a factorization

$X \to Y_1 \to Y$

of φ, then $\varphi(X) \hookrightarrow Y_1$ is a closed subspace.

c) For every $q \in \varphi(X)$ the canonical map

$$\widetilde{\varphi}'_q\colon \mathcal{O}_{\varphi(X),q} \to (\varphi_*\mathcal{O}_X)_q$$

associated to $\varphi'\colon X \to \varphi(X)$ is injective.

d) If X is reduced, then $\varphi(X)$ is reduced.

We call $\varphi(X) \hookrightarrow Y$ the *image* of X.

<u>Proof.</u> Since $\varphi_*\mathcal{O}_X$ is coherent by 1.17, $\mathcal{K}$ is coherent by 0.6. $|\varphi|(X)$ being closed in $|Y|$ we have for $q \in Y$

$$q \in |\varphi|(X) \Leftrightarrow (\varphi_*\mathcal{O}_X)_q \neq 0 \Leftrightarrow \mathcal{K}_q \neq \mathcal{O}_{Y,q} \Leftrightarrow q \in \operatorname{supp}(\mathcal{O}_Y/\mathcal{K}).$$

This proves a).

b) is an immediate consequence of the definition of $\mathcal{K}$ and 0.17. c) is clear and d) follows from the universal property.

<u>1.19.</u> A continuous map $\varphi\colon X \to Y$ of locally compact topological spaces is called *semi-proper*, if for every compact $K \subset Y$ there is a compact $K' \subset X$ such that $\varphi(K') = \varphi(X) \cap K$. Then clearly $\varphi(X)$ is closed in Y, but φ need not be a closed map. KUHLMANN [97] and WHITNEY [150] proved the following

<u>Semi-proper mapping theorem.</u> If $\varphi\colon X \to Y$ is a semiproper holomorphic map, then $|\varphi|(X)$ is an analytic subset in Y.

It should be noted that for a semi-proper holomorphic map the kernel of $\widetilde{\varphi}\colon \mathcal{O}_Y \to \varphi_*\mathcal{O}_X$ need not be coherent, hence the image of X cannot be provided with a canonical complex structure as in 1.18. An example of a non-proper but semi-proper map is given in Figure 2.

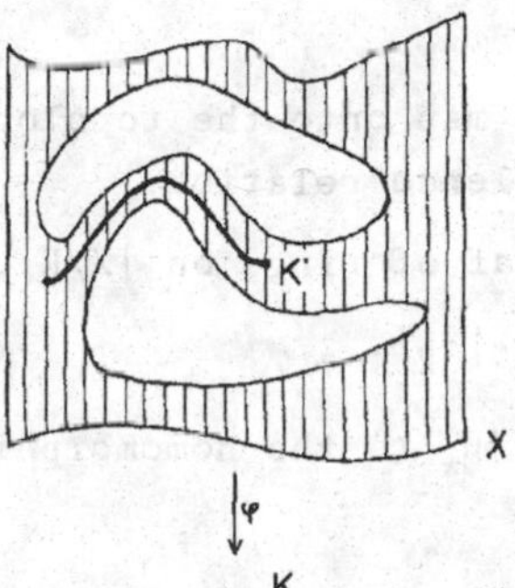

Figure 2

1.20. Let $\mathbb{C}$-ringed spaces R and X with a pair of morphisms ρ_1, ρ_2: R → X be given. Then a $\mathbb{C}$-ringed space X/R together with a morphism ψ: X → X/R such that $\psi \circ \rho_1 = \psi \circ \rho_2$ is called a *cokernel of the double arrow*

$$R \underset{\rho_2}{\overset{\rho_1}{\rightrightarrows}} X,$$

if the following universal property is satisfied:

If σ: X → S is any morphism of $\mathbb{C}$-ringed spaces such that $\sigma \circ \rho_1 = \sigma \circ \rho_2$, then there exists a uniquely determined morphism σ': X/R → S such that the diagram

$$\begin{array}{ccccc} & & & \overset{\sigma}{\nearrow} & S \\ & & & & \uparrow \sigma' \\ R & \underset{\rho_2}{\overset{\rho_1}{\rightrightarrows}} & X & \xrightarrow{\psi} & X/R \end{array}$$

commutes.

Proposition. For any double arrow $R \underset{\rho_2}{\overset{\rho_1}{\rightrightarrows}} X$ of $\mathbb{C}$-ringed spaces a cokernel ψ: X → X/R exists.

Of course, X/R need not be a complex space in case R and X are complex spaces. One essential obstruction is the fact that in general |X/R| is not hausdorff.

Proof. Call points $x, y \in X$ equivalent, if there is some $n \in \mathbb{N}$, $x_0, \ldots, x_n \in X$ and $r_1, \ldots, r_n \in R$ with the following properties:

$$x = x_0,\ x_n = y \text{ and } \begin{array}{l} \rho_1(r_i) = x_{i-1},\ \rho_2(r_i) = x_i \text{ or} \\ \rho_2(r_i) = x_{i-1},\ \rho_1(r_i) = x_i \text{ for } i = 1,..,n. \end{array}$$

Obviously this is the smallest equivalence relation such that x,y are equivalent, if $x = \rho_1(r)$ and $y = \rho_2(r)$ for some $r \in R$. Denote by

$$|\psi| : |X| \to |X/R|$$

the canonical continuous map onto the topological quotient with respect to the above equivalence relation.

In order to define a sheaf of rings on |X/R|, put

$$\xi := |\psi| \circ |\rho_1| = |\psi| \circ |\rho_2|.$$

For i = 1,2 we may apply $|\psi|_*$ to the homomorphisms

$$\tilde{\rho}_i : \mathcal{O}_X \to (\rho_i)_* \mathcal{O}_R$$

and this yields homomorphisms

$$|\psi|_* \mathcal{O}_X \overset{\alpha_1}{\underset{\alpha_2}{\rightrightarrows}} \xi_* \mathcal{O}_R.$$

We define

$$\mathcal{O}_{X/R} := \mathcal{K}er(\alpha_1 - \alpha_2) \subset |\psi|_* \mathcal{O}_X.$$

If

$$\tilde{\psi}: \mathcal{O}_{X/R} \to |\psi|_* \mathcal{O}_X$$

denotes the canonical injection, we obtain a morphism

$$\psi := (|\psi|, \tilde{\psi}): (|X|, \mathcal{O}_X) \to (|X/R|, \mathcal{O}_{X/R}) =: X/R.$$

Next we verify the not quite obvious fact that X/R is a local ringed space. For this purpose we consider a point $q \in X/R$ and $f \in \mathcal{O}_{X/R,q}$ represented by $F \in \mathcal{O}_{X/R}(V)$ for some open $V \subset X/R$ with $q \in V$. By definition of $\mathcal{O}_{X/R}$ we have $F \in \mathcal{O}_X(U)$ where $U := \psi^{-1}(V)$.
Given $r \in R$ and $x_1, x_2 \in U$ such that $x_1 = \rho_1(r)$, $x_2 = \rho_2(r)$, for F as above we obtain

$$F(x_1) = F(x_2)$$

by considering the local homomorphisms

$$\tilde{\rho}_{i,r}: \mathcal{O}_{X,x_i} \to \mathcal{O}_{R,r} \quad \text{for } i = 1,2.$$

The definition of our equivalence relation yields

$$F(x) = F(y)$$

for equivalent points $x,y \in U$ (shake the values through the chain!). Hence we may call $\mathcal{O}_{X/R}$ "the sheaf of R-invariant functions". In particular, we may assign a "value"

$$f(q) := F(x) \text{ for some } x \in \psi^{-1}(q)$$

to our given $f \in \mathcal{O}_{X/R,q}$.
Now it is easy to see that

$$\mathfrak{m} := \{f \in \mathcal{O}_{X/R,q}: f(q) = 0\}$$

is the unique maximal ideal. Assume $f \notin \mathfrak{m}$; since

$$\{x \in U: F(x) \neq 0\} = \{x \in U: F_x \in \mathcal{O}_{X,x} \text{ is a unit}\} \subset U$$

is open and the values of F are R-invariant we may choose V such that $F(x) \neq 0$ for any $x \in U$. Hence F is invertible in U and f is a unit.

Conversely, if $f \in \mathfrak{m}$, then $F(x) = 0$ for $x \in \psi^{-1}(q)$ and F_x is not invertible; hence f is not a unit.

Now it is obvious that ψ is a morphism of $\mathbb{C}$-ringed spaces. The proof of the universal property is straightforward and hence left to the pitiable reader.

<u>1.21.</u> It is a very difficult problem to find reasonable sufficient conditions for the cokernel of a double arrow of holomorphic maps to be a complex space.

If R and X are complex spaces, then a double arrow

$$R \overset{\rho_1}{\underset{\rho_2}{\rightrightarrows}} X$$

of holomorphic maps is called an *analytic equivalence relation*, if for every complex space Z the image of the map

$$\mathrm{Hol}(Z,R) \to \mathrm{Hol}(Z,X) \times \mathrm{Hol}(Z,X), \quad \psi \mapsto (\rho_1 \circ \psi, \rho_2 \circ \psi),$$

is an equivalence relation in the set Hol(Z,X).

A slightly more general notion is an *analytic pre-equivalence relation*, for which we refer to [85], [112].

We state one of the few existing results [85].

<u>Theorem.</u> Let X be a complex space and assume the double arrow

$$R \overset{\rho_1}{\underset{\rho_2}{\rightrightarrows}} X$$

belongs to an analytic pre-equivalence relation. If ρ_1 is flat (see 3.   ), ρ_2 is finite and if $|X/R|$ is hausdorff then the cokernel X/R is a complex space and the canonical holomorphic map $\psi: X \to X/R$ is open.

It is equally difficult, to find *fibre coproducts* in the category of complex spaces (see [126], [79]).

<u>1.22.</u> If $\varphi: X \to Y$ is a holomorphic map, we define

$$R_\varphi := X \times_Y X$$

and the canonical maps yield a double arrow

$$R_\varphi \overset{\pi_1}{\underset{\pi_2}{\rightrightarrows}} X.$$

By using GRAUERTs Coherence Theorem we show that its cokernel is a complex space in case φ is proper. At no extra cost we can even prove a little bit more (see [21], [78], [152]).

Theorem. Assume the diagram of complex spaces and holomorphic maps

$$R \underset{\rho_2}{\overset{\rho_1}{\rightrightarrows}} X \xrightarrow{\varphi} Y$$

has the following properties:

a) $\varphi \circ \rho_1 = \varphi \circ \rho_2$.

b) $|(\rho_1, \rho_2)(R)|$ generates the equivalence relation $|R_\varphi| \subset |X \times X|$ on $|X|$.

c) φ and $\varphi \circ \rho_1$ are proper.

Then there is a commutative diagram of complex spaces and holomorphic maps

$$\begin{array}{ccccc} R & \underset{\rho_2}{\overset{\rho_1}{\rightrightarrows}} & X & \xrightarrow{\psi} & X/R \\ & & \downarrow \varphi & \swarrow \tau & \downarrow \tau' \\ & & Y & \hookleftarrow & \varphi(X) \end{array}$$

such that

1) $\psi: X \to X/R$ is a cokernel of the given double arrow.

2) $|\tau'|$ is a homeomorphism.

In case $R = R_\varphi$, we obtain a *canonical factorization*

$$X \xrightarrow{\psi} X/R_\varphi \xrightarrow{\tau'} \varphi(X) \hookrightarrow Y$$

of the holomorphic map φ. If, in addition, φ is flat (see 3.11), then τ' is biholomorphic.

Proof. According to 1.20 we have a cokernel

$$\psi: X \to X/R$$

in the category of $\mathbb{C}$-ringed spaces. It is well known from topology (see e.g. [12]) that X/R is locally compact and ψ is proper, since φ was proper. Furthermore the canonical map

$$|\tau'|: |X/R| \to |\varphi(X)|$$

is a homeomorphism. Hence we may consider

$$|X/R| \subset |Y|$$

as a closed subspace and we have $|\varphi| = |\psi|$.

The holomorphic maps φ and

$\xi := \varphi \circ \rho_1 : R \to Y$

are proper; hence the $\mathcal{O}_Y$-modules in the exact sequence

$$\xi_* \mathcal{O}_R \underset{\alpha_2}{\overset{\alpha_1}{\longleftleftarrows}} \varphi_* \mathcal{O}_X \hookleftarrow \mathcal{O}_{X/R}$$

constructed as in 1.20 are $\mathcal{O}_Y$-coherent. We consider the analytic spectrum

χ: Specan $\mathcal{O}_{X/R} \to Y$.

Since $\mathcal{O}_{X/R}$ is a sheaf of local algebras over $|\varphi(X)|$, $|\chi|$ is injective by 1.15 and its image is equal to $|\varphi(X)|$. This yields an isomorphism

Specan $\mathcal{O}_{X/R} \to X/R$

of $\mathbb{C}$-ringed spaces, hence X/R is a complex space.

In order to complete the promised diagram it suffices to consider the diagram

$$\begin{array}{ccc} \varphi_* \mathcal{O}_X & \hookleftarrow & \mathcal{O}_{X/R} \\ \uparrow{\scriptstyle \tilde{\varphi}} & \nearrow & \uparrow \\ \mathcal{O}_Y & \longrightarrow & \mathcal{O}_Y / \mathit{Ker}\, \tilde{\varphi}. \end{array}$$

Obviously τ' is biholomorphic if and only if

$\tilde{\varphi}\, \mathcal{O}_Y = \mathcal{O}_{X/R}$.

This is the case if φ is flat (for details we refer to [85]).

<u>1.23. Lemma.</u> Let $\varphi: X \to Y$ be a proper holomorphic map such that the canonical map $\tilde{\varphi}: \mathcal{O}_Y \to \varphi_* \mathcal{O}_X$ is an isomorphism. Then all the fibres of φ are connected and non-empty.

<u>Proof.</u> Assume the fibre of $q \in Y$ is not connected. Since φ is proper we may find an open neighbourhood V of q such that $\varphi^{-1}(V)$ is not connected. But then the canonical map

$\mathcal{O}_{Y,q} \to (\varphi_* \mathcal{O}_X)_q$

cannot be surjective.

If the fibre of q were empty this map could not be injective.

<u>1.24.</u> Now we can give a short proof (due to GROTHENDIECK [53]) of a factorization theorem for proper holomorphic maps [138], [21], [119].

<u>Theorem.</u> Let $\varphi: X \to Y$ be a proper holomorphic map. Then there is a

commutative diagram

$$\begin{array}{ccccc} X & & \overset{\sigma}{\longrightarrow} & & \\ \downarrow{\scriptstyle\varphi} & & & \searrow & \\ Y & \underset{\tau}{\longleftarrow} & X/R_{\varphi} & \underset{\chi}{\longleftarrow} & Z \end{array}$$

of complex spaces and holomorphic maps with the following properties:

1) τ is the canonical map of 1.22

2) χ and $\tau\cdot\chi$ are finite

3) σ is proper, surjective, has connected fibres and the canonical map

$$\tilde{\sigma}: \mathcal{O}_Z \to \sigma_*\mathcal{O}_X$$

is an isomorphism.

Z may be considered as the *space of level sets of* φ.

The above diagram is called the *Stein factorization* of φ.

Proof. Recall that X/R_φ = Specan A, where $A \subset \varphi_*\mathcal{O}_X$ is the kernel of

$$\alpha_1-\alpha_2: \varphi_*\mathcal{O}_X \to \xi_*\mathcal{O}_{R_\varphi}$$

(notations as in 1.22). Since φ was proper we obtain a complex space

$$Z := \text{Specan } \varphi_*\mathcal{O}_X$$

with a finite holomorphic map $\zeta: Z \to Y$ and an isomorphism

$$\varphi_*\mathcal{O}_X \xrightarrow{\sim} \zeta_*\mathcal{O}_Z$$

(see 1.10 and 1.14). Now the bijective map

$$\text{Hol}_Y(Z, X/R_\varphi) \to \text{Hom}_{\mathcal{O}_Y\text{-alg}}(A, \zeta_*\mathcal{O}_Z)$$

determines $\chi: Z \to X/R_\varphi$ as the inverse image of the homomorphism

$$A \hookrightarrow \varphi_*\mathcal{O}_X \xrightarrow{\sim} \zeta_*\mathcal{O}_Z.$$

In the same way the identity of $\varphi_*\mathcal{O}_X$ determines $\sigma: X \to Z$ and obviously everything commutes. Applying ζ_* to

$$\tilde{\sigma}: \mathcal{O}_Z \to \sigma_*\mathcal{O}_X$$

yields

$$\zeta_*\mathcal{O}_Z \to \varphi_*\mathcal{O}_X.$$

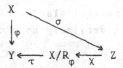

Since this is an isomorphism 1.14 implies that $\tilde{\sigma}$ is an isomorphism. By 1.23 the fibres of σ are connected and non empty. Since φ is proper and ζ is finite we conclude that σ is proper and χ is finite.

1.25. As an important application of the results on equivalence relations we show how to associate a Stein space to an arbitrary holomorphically convex space [116], [21], [152].
For a relative version of this result see [90].

Theorem. Let X be a holomorphically convex complex space. Then there exists a holomorphic map $\sigma: X \to X_{pet}$ with the following properties:

1) X_{pet} is a Stein space
2) σ is proper with connected fibres
3) the canonical maps $\tilde{\sigma}: \mathcal{O}_{X_{pet}} \to \sigma_* \mathcal{O}_X$ and $\mathcal{O}_{X_{pet}}(X_{pet}) \to \mathcal{O}_X(X)$ are isomorphisms.
4) If $\varphi: X \to Y$ is any holomorphic map into a Stein space Y then there is a uniquely determined holomorphic map $\psi: X_{pet} \to Y$ such that the diagram

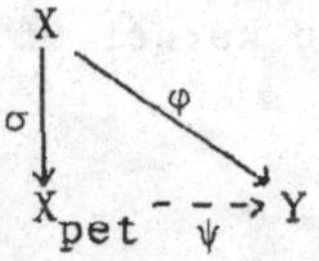

commutes. In particular X_{pet} is uniquely determined by X.

We call $\sigma: X \to X_{pet}$ the *petrification* of X.

Proof. If $A := \mathcal{O}_X(X)$ we consider the equivalence relation $|R| \subset |X \times X|$ defined by

$$x \sim x' \text{ if } f(x) = f(x') \text{ for all } f \in A.$$

Denote by

$$\zeta: |X| \to |Z|$$

the canonical map on the topological quotient. If $K \subset X$ is compact, then

$$\zeta^{-1}(\zeta(K)) \subset \hat{K} = \{x \in X: |f(x)| \leq \sup|f(K)| \text{ for all } f \in A\}$$

is a closed subset. Since X is holomorphically convex, $\hat{K}$ is compact and hence $\zeta^{-1}(\zeta(K))$ is compact. This implies that $|Z|$ is locally compact and ζ is proper (see [12]).

In order to obtain a ringed structure on $|Z|$, we consider the sheaf of ideals $\mathcal{I} \subset \mathcal{O}_{X \times X}$ generated by

$$\{f \otimes 1 - 1 \otimes f: f \in A\} \subset \mathcal{O}_{X \times X}(X \times X).$$

$\mathcal{I}$ is coherent by 0.40; denote by $R \hookrightarrow X \times X$ the corresponding closed complex subspace. $|Z|$ is the underlying topological space of the cokernel of the resulting double arrow; we denote this cokernel by

$$\zeta : X \to Z = (|Z|, \mathcal{O}_Z).$$

Next we show that Z is a complex space. If $V \subset Z$ is relatively compact, then

$$U := \zeta^{-1}(V) \subset X \text{ and } U \times U \subset X \times X$$

are relatively compact. Hence there is a finite subset $\{f_1, \ldots, f_n\} \subset A$ such that

$$\{f_i \otimes 1 - 1 \otimes f_i : i = 1, \ldots, n\}$$

generate $\mathcal{I}$ on $U \times U$. Hence the holomorphic map

$$\varphi = (f_1, \ldots, f_n) : U \to \mathbb{C}^n$$

(see 1.10)

has the same fibres as $\zeta|U$. A simple topological argument shows that we may find an open $W \subset \mathbb{C}^n$ such that there is a factorization

$$U \xrightarrow{\varphi'} W \subset \mathbb{C}^n$$

of φ, where φ' is proper. Obviously R and $R_{\varphi'}$ coincide on $U \times U$. Hence we have an isomorphism

$$U/R_{\varphi'} \to (V, \mathcal{O}_Z|V)$$

and 1.22 implies that Z is a complex space.
From the definition of $\mathcal{O}_Z$ we obtain (see 1.20)

$$\mathcal{O}_Z(Z) = \mathcal{O}_X(X) = A.$$

This implies $\hat{L} = \zeta(\hat{K})$ for $L \subset Z$ compact and $K := \zeta^{-1}(L)$. Hence Z is holomorphically convex. Since Z is holomorphically separable by definition we have proved that Z is Stein.

Let

$$X \xrightarrow{\sigma} X_{pet} \xrightarrow{\xi} Z$$

be a Stein factorization of ζ, where ξ is finite. X_{pet} is Stein by 1.13; the assertions 2) and 3) follow from 1.24.

In order to prove the universal property of 4) we consider the homomorphism

$$\mathcal{O}_{X_{pet}}(X_{pet}) = \mathcal{O}_X(X) \leftarrow \mathcal{O}_Y(Y)$$

determined by φ. By a theorem on Stein algebras [33] it uniquely determines

$$X_{pet} \to Y$$

as desired.

Corollary. Let X be a holomorphically convex complex space such that every compact analytic subset of X is finite. Then X is Stein.

Proof. Since the petrification

$$\sigma: X \to X_{pet}$$

is proper with connected fibres, $|\sigma|$ must be a homeomorphism. Hence X is Stein by 1.13. (By using the universal property, we even see that σ is biholomorphic.)

1.26. Finally we consider the situation where an equivalence relation

$$T \subset |X \times X|$$

on the underlying topological space of a complex space X is given. Denote by

$$|\varphi|: |X| \to |Y|$$

the topological quotient map. Then we define a subsheaf $A \subset \varphi_* \mathcal{O}_X$ of T-*invariant functions* by

$$A(V) = \{f \in \mathcal{O}_X(U): f(x) = f(x') \text{ if } x \sim x'\}$$

for $V \subset |Y|$ open and $U := |\varphi|^{-1}(V)$. It is easy to see that A is a sheaf of local $\mathbb{C}$-algebras on $|Y|$, hence

$$Y := (|Y|, A)$$

is a $\mathbb{C}$-ringed space. If $\tilde{\varphi}: A \to \varphi_* \mathcal{O}_X$ denotes the injection, we obtain a morphism of $\mathbb{C}$-ringed spaces

$$\varphi = (|\varphi|, \tilde{\varphi}): X \to Y.$$

Recall that the equivalence relation T is called *proper*, if

a) $|Y|$ is locally compact,

b) $|\varphi|$ is proper.

Now we can formulate a result due to H. CARTAN [21] (see also [78], [152]).

Theorem. In the situation as above, assume T is proper. Then the following conditions are equivalent:

i) Y is a complex space.

ii) Y is locally A-separable.

Proof. i) ⇒ ii) is obvious from the definition of a complex space. To show ii) ⇒ i) assume every $q \in Y$ has an open neighbourhood V such that the map

$$G: V \to \mathbb{C}^A, \; y \mapsto (f(y))_{f \in A}, \text{ with } A := A(V)$$

is injective. Since the elements of A are holomorphic functions on $U := \varphi^{-1}(V)$ we obtain a commutative diagram

$$\begin{array}{ccc} U & \xrightarrow{F} & \mathbb{C}^A \\ \varphi \downarrow & \nearrow G & \\ V & & \end{array}$$

of continuous maps. Since G is injective, we have

$$R_F = T \text{ on } U,$$

if R_F denotes the topological equivalence relation determined by F. Consider the sheaf of ideals $I \subset \mathcal{O}_{U \times U}$ generated by

$$\{f \otimes 1 - 1 \otimes f : f \in A\}.$$

Since T is proper we may find an open $V_o \subset Y$ such that

$$q \in V_o \subset\subset V \text{ and } U_o := \varphi^{-1}(V_o) \subset\subset U.$$

Since I is coherent (0.40) we may find $f_1, \ldots, f_n \in A$ such that I is generated by

$$\{f_i \otimes 1 - 1 \otimes f_i : i = 1, \ldots, n\} \text{ on } U.$$

If we define $F_o = G_o = (f_1, \ldots, f_n)$, we obtain a commutative diagram

$$\begin{array}{ccc} U_o & \xrightarrow{F_o} & \mathbb{C}^n \\ \varphi | U_o \downarrow & \nearrow G_o & \\ V_o, & & \end{array}$$

where F_o is holomorphic and G_o is injective. For topological reasons we may shrink V_o (and hence U_o) and find an open neighbourhood W of $G_o(q)$ in $\mathbb{C}^n$ with a factorization

$$U_o \xrightarrow{F'_o} W \subset \mathbb{C}^n$$

of F_o, where F'_o is proper (see 1.10).

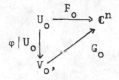

On U_o the given equivalence relation T coincides with $|R_{F'_o}|$. Furthermore we define

$$R := (R_{F'_o})_{red} \hookrightarrow U_o \times U_o;$$

from the definition of A it is obvious, that there is an isomorphism

$$U_o/R \to (V_o, A|V_o).$$

Since the cokernel U_o/R is a complex space by 1.22, the theorem is proved.

★

Chapter 2. Differential Calculus

In this chapter we first develop the elementary concepts of "differential calculus" on complex spaces such as tangent vectors, vector fields and differential forms. They are the principal tool for the study of singularities. Our aim is the proof of several "differential regularity criterions" which may be considered as a generalization of the classical implicit function theorem.

2.1. Let X be a complex space. For $p \in X$ a $\mathbb{C}$-linear map

$$\delta_p : \mathcal{O}_{X,p} \to \mathbb{C}$$

is called a *tangent vector to* X *at* p, (or a *derivation* of $\mathcal{O}_{X,p}$), if it satisfies one of the following equivalent conditions:

i) $\delta_p(f\cdot g) = g(p)\cdot\delta_p(f) + f(p)\cdot\delta_p(g)$ for all $f,g \in \mathcal{O}_{X,p}$

ii) $\delta_p(\mathfrak{m}_{X,p}^2) = 0$ and $\delta_p(\mathbb{C}\cdot 1) = 0$.

We prove the equivalence of i) and ii).

i) $\Rightarrow$ ii) is obvious: If

$f,g \in \mathfrak{m}_{X,p}$, then $\delta_p(f\cdot g) = 0$ and
$\delta_p(1) = \delta_p(1\cdot 1) = \delta_p(1) + \delta_p(1) = 0$.

In order to prove ii) $\Rightarrow$ i) we use the decomposition

$$\mathcal{O}_{X,p} = \mathbb{C} \oplus \mathfrak{m}_{X,p}, \quad f = f(p) + \overline{f}.$$

Then

$$\begin{aligned}\delta_p(f\cdot g) &= \delta_p((f(p) + \overline{f})(g(p) + \overline{g}))\\ &= \delta_p(f(p)g(p)) + \delta_p(\overline{f}\cdot\overline{g}) + g(p)\delta_p(\overline{f}) + f(p)\delta_p(\overline{g})\\ &= g(p)\delta_p(f) + f(p)\delta_p(g).\end{aligned}$$

Obviously the set of all tangent vectors to X at p is a $\mathbb{C}$-vector space. We denote it by

$$T_pX$$

and call it the *tangent space of* X *at* p.

If $\varphi: X \to Y$ is a holomorphic map with $\varphi(p) = q$, and if

$$\tilde{\varphi}_p : \mathcal{O}_{Y,q} \to \mathcal{O}_{X,p}$$

denotes the canonical homomorphism (see 0.3), then

$$T_p\varphi : T_pX \to T_qY,\ \delta_p \mapsto \delta_p \circ \tilde{\varphi}_p,$$

is a homomorphism of $\mathbb{C}$-vector spaces. We call it the *Jacobian map of* φ *at* p.

If δ_p is a tangent vector to X at p, then the restriction of δ_p to $\mathfrak{m}_{X,p}$ induces a $\mathbb{C}$-linear map

$$\mathfrak{m}_{X,p}/\mathfrak{m}^2_{X,p} \to \mathbb{C}.$$

In this way we obtain a map

$$T_pX \to (\mathfrak{m}_{X,p}/\mathfrak{m}^2_{X,p})^*$$

which turns out to be an isomorphism of $\mathbb{C}$-vector spaces. This may easily be seen by using condition ii) of the definition of tangent vectors.

2.2. Next we give a useful description of tangent spaces by means of local coordinates.

Consider an open subset $W \subset \mathbb{C}^m$ with coordinate functions $w_1,\dots,w_m$. For $p \in X$ and $i = 1,\dots,m$ we have tangent vectors

$$\frac{\partial}{\partial w_i}(p) : \mathcal{O}_{W,p} \to \mathbb{C},\ f \mapsto \frac{\partial F}{\partial w_i}(p),$$

where F represents f in some neighbourhood of p.
It is well known that the map

$$\mathbb{C}^m \to T_pW,\ (s_1,\dots,s_m) \mapsto s_1 \frac{\partial}{\partial w_1}(p) + \dots + s_m \frac{\partial}{\partial w_m}(p),$$

is an isomorphism of $\mathbb{C}$-vector spaces.
If X is a closed complex subspace of W with canonical injection

$$\iota : X \hookrightarrow W,$$

defined by the coherent ideal $\mathcal{I} \subset \mathcal{O}_W$, we have an exact sequence

$$0 \to \mathcal{I}_p \to \mathcal{O}_{W,p} \to \mathcal{O}_{X,p} \to 0$$

for any $p \in X$. This immediately implies that

$$T_p\iota : T_pX \to T_pW$$

is injective and its image is equal to

$$\{\delta_p \in T_pW : \delta_p(\mathcal{I}_p) = 0\}.$$

In terms of local coordinates this means that T_pX is isomorphic to

$\{(s_1,\ldots,s_m) \in \mathbb{C}^m: s_1 \frac{\partial f}{\partial w_1}(p) + \ldots + s_m \frac{\partial f}{\partial w_m}(p) = 0 \text{ for all } f \in I_p\}.$

Obviously it is sufficient to check this condition for a set of generators of I_p.

2.3. As we have just seen the dimension of the tangent space at a point is always finite. We want to give a geometric interpretation of this number.

If p is a point of a complex space X, we define the *embedding dimension of* X *at* p, denoted by

$$\operatorname{emdim}_p X,$$

as the minimal $n \in \mathbb{N}$ such that there exists an open neighbourhood U of p and a holomorphic immersion $U \to \mathbb{C}^n$ (see 0.23).

Proposition. For every point p of a complex space X we have

$$\operatorname{emdim}_p X = \dim_{\mathbb{C}} T_p X.$$

Proof. Put $m := \dim_{\mathbb{C}} T_p X$. As we have seen in 2.2, the Jacobian map of an immersion is injective, hence $\operatorname{emdim}_p X \geq m$.

To prove the opposite inequality we use the isomorphism

$$T_p X \to (\mathfrak{m}_{X,p}/\mathfrak{m}_{X,p}^2)^*$$

of 2.1. Since

$$m = \dim_{\mathbb{C}}(\mathfrak{m}_{X,p}/\mathfrak{m}_{X,p}^2)^* = \dim_{\mathbb{C}}(\mathfrak{m}_{X,p}/\mathfrak{m}_{X,p}^2)$$

the lemma of NAKAYAMA yields generators $f_1,\ldots,f_m \in \mathcal{O}_{X,p}$ of $\mathfrak{m}_{X,p}$. If we choose an open neighbourhood U of p and representatives $\varphi_1,\ldots,\varphi_m \in \mathcal{O}_X(U)$ of $f_1,\ldots,f_m$, then

$$(\varphi_1,\ldots,\varphi_m): U \to \mathbb{C}^m$$

is an immersion at p by 0.23.

2.4. Proposition. Let $\varphi: X \to Y$ be a holomorphic map, $p \in X$ and $q := \varphi(p)$. Then φ is an immersion at p if and only if the Jacobian map

$$T_p\varphi: T_p X \to T_p Y$$

is injective.

If in addition X is non-singular at p (see 0.14) then φ is biholomorphic at p if and only if the Jacobian map $T_p\varphi$ is bijective.

Proof. The necessity of the conditions is obvious. In order to prove the sufficiency we may suppose that we have a commutative diagram

$$\begin{array}{ccccc} & \mathbb{C}^m & & \mathbb{C}^n & \\ & \cup & & \cup & \\ & W & \overset{\Phi}{\dashrightarrow} & Z & \\ & \uparrow & & \uparrow & \\ X \supset & U & \xrightarrow{\varphi|U} & V & \subset Y, \\ & \in & & \in & \\ & p & & q & \end{array}$$

where m := $\mathrm{emdim}_p X$, n := $\mathrm{emdim}_q Y$ and Φ is choosen according to Corollary 2 of 0.22. By 2.2 we have $T_pX = T_pW$ and $T_qY = T_qZ$, hence $T_p\varphi = T_p\Phi$. The classical implicit function theorem applied to Φ shows that Φ is an immersion at p and hence $\tilde{\Phi}_p$ is surjective. If we look at the diagram

$$\begin{array}{ccc} \mathcal{O}_{W,p} & \xleftarrow{\tilde{\Phi}_p} & \mathcal{O}_{Z,q} \\ \downarrow & & \downarrow \\ \mathcal{O}_{X,p} & \xleftarrow{\tilde{\varphi}_p} & \mathcal{O}_{Y,q} \end{array}$$

we see that $\tilde{\varphi}_p$ is surjective and φ is an immersion at p by 0.23. In case X is non-singular at p we may suppose U = W. If $T_p\varphi = T_p\Phi$ is bijective we have m = n. As we have just seen, φ is an immersion at p, hence we may suppose that there is a factorization

$$W \xrightarrow{\varphi'} V' \hookrightarrow V$$

of $\varphi|W$, where φ' is biholomorphic. Since

$$\mathrm{emdim}_q V' = m,$$

we may suppose V' = V = Z, hence $\varphi' = \varphi|U = \Phi$.

<u>2.5.</u> So far we have only considered tangent vectors at a point of a complex space. If we want to define vector fields and differential forms the tangent vector spaces have to be glued together to a linear fibre space $TX \to X$.
For the underlying set of TX we simply take the disjoint union

$$\bigcup_{p \in X} T_pX.$$

In order to define a complex structure on TX, we first assume that

$$X \hookrightarrow W \subset \mathbb{C}^m$$

is a closed complex subspace of an open subset W of $\mathbb{C}^m$, defined by a sheaf of ideals $\mathcal{I} \subset \mathcal{O}_W$ which is generated by

$$f_1,\ldots,f_r \in \mathcal{O}_W(W).$$

If $(w_1,\ldots,w_m,s_1,\ldots,s_m)$ are coordinates in $W\times\mathbb{C}^m$ we define

$$TX \hookrightarrow W\times\mathbb{C}^m$$

as the closed complex subspace generated by

$$f_1,\ldots,f_r \text{ and } \frac{\partial f_\rho}{\partial w_1} s_1 +\ldots+ \frac{\partial f_\rho}{\partial w_m} s_m \text{ for } \rho = 1,\ldots,r,$$

where f_ρ and $\frac{\partial f_\rho}{\partial w_\mu}$ are considered as holomorphic functions on $W\times\mathbb{C}^m$ via the canonical projection $W\times\mathbb{C}^m \to W$. In fact

$$TX \hookrightarrow X\times\mathbb{C}^m$$

is the subspace generated by the linear forms on $X\times\mathbb{C}^m$

$$f_{\rho,1}s_1 +\ldots+ f_{\rho,m}s_m \text{ for } \rho = 1,\ldots,r,$$

where $f_{\rho,\mu} \in \mathcal{O}_X(X)$ is the residue class of $\frac{\partial f_\rho}{\partial w_\mu} \in \mathcal{O}_W(W)$.
Hence TX is a linear subspace of $X\times\mathbb{C}^m$ and the projection of $X\times\mathbb{C}^m$ induces

$$\pi: TX \to X.$$

The coordinate description of 2.2 immediately implies

$$T_pX = (TX)_p.$$

It is an easy exercise to show that this construction does not depend on the choices made. Hence for an arbitrary complex space X the local pieces of the tangent space may be glued together according to 0.24 and we obtain the *tangent space of* X, denoted by TX, which is a linear space over X.
One should always keep in mind that this canonical construction yields a complex structure on TX which, in general, is not reduced even if X was reduced (compare [114]).

2.6. Next we have to glue together the Jacobian maps. More precisely, given a holomorphic map

$$\varphi: X \to Y$$

we have to define a homomorphism of linear spaces over X

$$T\varphi: TX \to \varphi^*TY$$

(see 1.7). The problem being local with respect to X and Y, we may suppose that we have a commutative diagram

$$\begin{array}{ccc} \mathbb{C}^m & & \mathbb{C}^n \\ \cup & & \cup \\ W & \xrightarrow{\Phi} & Z \\ \uparrow & & \uparrow \\ X & \xrightarrow{\varphi} & Y, \end{array}$$

i.e. φ is generated by Φ. We may further assume that the ideal

$I \subset \mathcal{O}_W$ of X is generated by $f_1,\ldots,f_k \in \mathcal{O}_W(W)$ and
$J \subset \mathcal{O}_Z$ of Y is generated by $g_1,\ldots,g_l \in \mathcal{O}_Z(Z)$.

In order to define $T\varphi$ we consider the diagram

$$\begin{array}{ccccc} TW & & \Phi^*TZ & & TZ \\ \| & & \| & & \| \\ W\times\mathbb{C}^m & \xrightarrow{T\Phi} & W\times\mathbb{C}^n & \xrightarrow{\Phi\times id} & Z\times\mathbb{C}^n \\ \uparrow & & \uparrow & & \uparrow \\ TX & \dashrightarrow^{T\varphi} & \varphi^*TY & \xrightarrow{\varphi\times id} & TY, \end{array}$$

where $T\Phi$ is the classical Jacobian map, defined by

$$(w,s_1,\ldots,s_m) \mapsto \left(w, \sum_{\mu=1}^{m} \frac{\partial\Phi_1}{\Phi w_\mu} s_\mu,\ldots, \sum_{\mu=1}^{m} \frac{\partial\Phi_n}{\partial w_\mu} s_\mu\right),$$

and we have to show that $T\Phi$ admits a restriction $T\varphi$ (see 0.17). According to 2.5, $TY \hookrightarrow Z\times\mathbb{C}^n$ is the subspace generated by

$$g_1,\ldots,g_l \text{ and } \frac{\partial g_\lambda}{\partial z_1} t_1 +\ldots+ \frac{\partial g_\lambda}{\partial z_n} t_n \text{ for } \lambda = 1,\ldots,l,$$

where $z_1,\ldots,z_n$ are coordinates in Z and $t_1,\ldots,t_n$ are coordinates in $\mathbb{C}^n$. Consequently $\varphi^*TY \hookrightarrow W\times\mathbb{C}^n$ is the subspace generated by

$$f_1,\ldots,f_k \text{ and } \left(\frac{\partial g_\lambda}{\partial z_1}\circ\Phi\right) t_1 +\ldots+ \left(\frac{\partial g_\lambda}{\partial z_n}\circ\Phi\right) t_n \text{ for } \lambda = 1,\ldots,l. \qquad (*)$$

To apply the criterion of 0.17 we have to substitute

$$t_\nu = \frac{\partial\Phi_\nu}{\partial w_1} s_1 +\ldots+ \frac{\partial\Phi_\nu}{\partial w_m} s_m$$

in the linear forms of (*). By an elementary computation this yields

$$\frac{\partial(g_\lambda\circ\Phi)}{\partial w_1} s_1 +\ldots+ \frac{\partial(g_\lambda\circ\Phi)}{\partial w_m} s_m. \qquad (**)$$

Since φ has been the restriction of Φ, the function $g_\lambda \circ \Phi$ is a section in I over W. This implies that (**) is a section in the ideal of $TX \hookrightarrow W \times \mathbb{C}^m$ if we shrink W sufficiently and finally we have seen that there is a restriction

$$T\varphi: TX \to \varphi^* TY$$

of $T\Phi$. We call $T\varphi$ the *Jacobian map of* φ.
We leave it as an exercise to the reader to show that for any $p \in X$

$$(T\varphi)_p = T_p\varphi.$$

As an immediate consequence of 2.4 and the above construction of the Jacobian map we obtain the

Remark. A holomorphic map $\varphi: X \to Y$ is an immersion if and only if the Jacobian map

$$T\varphi: TX \to \varphi^* TY$$

is a monomorphism.

2.7. If $\varphi: X \to Y$ is a holomorphic map, we define

$$T(X/Y) := \operatorname{Ker} T\varphi \hookrightarrow TX$$

and call it the *tangent space of* φ (or *the tangent space of* X *over* Y). By definition it is a linear space over X which is a linear subspace of TX. Geometrically, T(X/Y) consists of all tangent vectors "in the fibre direction". More precisely, we state the

Remark. Given a holomorphic map $\varphi: X \to Y$ and $q \in Y$, denote by $\iota: X_q \hookrightarrow X$ the canonical injection. Then

$$T(X_q) = \iota^* T(X/Y).$$

In order to prove this, we use the following coordinate description of T(X/Y). In the situation and with the notations of 2.6,

$$T(X/Y) \hookrightarrow W \times \mathbb{C}^m$$

is the complex subspace generated by the holomorphic functions in $W \times \mathbb{C}^m$

$$f_1, \ldots, f_k,$$

$$\frac{\partial f_\varkappa}{\partial w_1} s_1 + \ldots + \frac{\partial f_\varkappa}{\partial w_m} s_m, \text{ for } \varkappa = 1, \ldots, k, \qquad (*)$$

$$\frac{\partial \Phi_\nu}{\partial w_1} s_1 + \ldots + \frac{\partial \Phi_\nu}{\partial w_m} s_m, \text{ for } \nu = 1, \ldots, n. \qquad (**)$$

This follows immediately from the definition of $T\varphi$ in 2.6.

To prove our remark we may suppose we are in the situation of 2.6, since the question is local with respect to X. We may further assume $q = o \in Z$. Then by 0.27

$$X_q \hookrightarrow W$$

is the subspace generated by $f_1,\ldots,f_k$ and $\Phi_1,\ldots,\Phi_n$. Consequently

$$T(X_q) \hookrightarrow W\times\mathbb{C}^m$$

is the subspace generated by $f_1,\ldots,f_k$ and $\Phi_1,\ldots,\Phi_n$ together with the functions (*) and (**). Since

$$\iota^*T(X/Y) \hookrightarrow T(X/Y)$$

is just the subspace generated by the residue classes of $\Phi_1,\ldots,\Phi_n$, the remark is verified.

2.8. As an easy consequence of the definition of the tangent space given in 2.5 and the construction of the direct product of complex spaces in 0.28 we obtain the

Lemma. Let X and Y be complex space and denote by $\pi_X: X\times Y \to X$ and $\pi_Y: X\times Y \to Y$ the canonical projections. Then

$$T(X\times Y) = \pi_X^*TX \times_{X\times Y} \pi_Y^*TY.$$

2.9. If X is a complex space, any homomorphism

$$\omega: TX \to X\times\mathbb{C}$$

of linear spaces over X is called a *Pfaffian form* (or *differential form of type* (1,0)). By

$$\Omega_X := \mathcal{H}om_X(TX, X\times\mathbb{C})$$

we denote the *sheaf of Pfaffian forms*.

If $\varphi: X \to Y$ is a holomorphic map, we obtain a homomorphism

$$D\varphi: \varphi^*\Omega_Y \to \Omega_X$$

as follows: Given $U \subset X$ open and

$$\omega \in \varphi^*\Omega_Y(U) = \mathrm{Hom}(\varphi^*TY_U, U\times\mathbb{C}),$$

we define $D\varphi(U)\omega$ as the composite map

$$TX_U \xrightarrow{T\varphi} \varphi^*TY_U \xrightarrow{\omega} U\times\mathbb{C},$$

where $T\varphi$ denotes the Jacobian map defined in 2.6.

In the case of a direct product we have, with the notations of 2.8,

$$\Omega_{X\times Y} = \pi_X^* \Omega_X \oplus \pi_Y^* \Omega_Y.$$

If X is a complex space, any holomorphic section

$$\delta: X \to TX$$

(i.e. a holomorphic map such that $\pi \circ \delta = id_X$, where $\pi: TX \to X$ denotes the canonical projection) is called a *vector field* on X. If T_X denotes the *sheaf of vector fields*, we obviously have

$$T_X = Hom_X(X\times\mathbb{C}, TX).$$

Of course vector fields, in general, cannot be transported via a holomorphic map.

2.10. In order to establish a fundamental relation between Pfaffian forms and vector fields, we recall a well known definition.

If R is a $\mathbb{C}$-algebra and M is an R-module, then a map

$$\delta: R \to M$$

is called a *derivation,* if

a) δ is $\mathbb{C}$-linear

b) $\delta(f\cdot g) = g\cdot\delta(f) + f\cdot\delta(g)$.

If X is a complex space and F is an $\mathcal{O}_X$-module, then

$$d: \mathcal{O}_X \to F$$

is called a *derivation,* if $d(U): \mathcal{O}_X(U) \to F(U)$ is a derivation for any open $U \subset X$. By

$$\mathrm{Der}\ \mathcal{O}_X$$

we denote the $\mathcal{O}_X(X)$-module of derivations

$$\mathcal{O}_X \to \mathcal{O}_X.$$

Theorem. For every complex space X there is a canonical derivation

$$d_X: \mathcal{O}_X \to \Omega_X$$

such that the map

$$\mathrm{Hom}(\Omega_X, \mathcal{O}_X) \to \mathrm{Der}\ \mathcal{O}_X,\ \xi \mapsto \xi\circ d_X,$$

is an isomorphism of $\mathcal{O}_X(X)$-modules, i.e. for any derivation $d \in \mathrm{Der}\ \mathcal{O}_X$ there is a uniquely determined homomorphism $\xi \in \mathrm{Hom}(\Omega_X, \mathcal{O}_X)$ such that the diagram

$$\begin{array}{ccc} \mathcal{O}_X & & \\ {\scriptstyle d_X}\downarrow & \searrow{\scriptstyle d} & \\ \Omega_X & \overset{}{\underset{\xi}{\dashrightarrow}} & \mathcal{O}_X \end{array}$$

commutes.

We only sketch the <u>proof</u>;for details the reader may consult [50]. First we consider an open subset $W \subset \mathbb{C}^m$ with coordinates $w_1,\ldots,w_m$. Then we have the Pfaffian forms

$$dw_i : TW = W\times\mathbb{C}^m \to W\times\mathbb{C},\ (w,s_1,\ldots,s_m) \mapsto (w,s_i),$$

for $i = 1,\ldots,m$ and an isomorphism

$\mathcal{O}_W^m \to \Omega_W$, given by

$$\mathcal{O}_W(U)^m \to \Omega_W(U),\ (\alpha_1,\ldots,\alpha_m) \mapsto \alpha_1 dw_1 + \ldots + \alpha_m dw_m$$

for $U \subset W$ open. Now we define

$d_W\colon \mathcal{O}_W \to \Omega_W$ by

$$\mathcal{O}_W(U) \to \Omega_W(U),\ F \mapsto \frac{\partial F}{\partial w_1}\, dw_1 + \ldots + \frac{\partial F}{\partial w_m}\, dw_m,$$

for $U \subset W$ open. In order to prove the universal property consider the diagram

$$\begin{array}{ccc} \mathcal{O}_W(U) & & \\ {\scriptstyle d_W(U)}\downarrow & \searrow{\scriptstyle d(U)} & \\ \Omega_W(U) & \underset{\xi(U)}{\dashrightarrow} & \mathcal{O}_W(U). \end{array}$$

Since Der $\mathcal{O}_W(U)$ is generated by $\frac{\partial}{\partial w_1},\ldots,\frac{\partial}{\partial w_m}$ we may assume

$$d(U) = \frac{\partial}{\partial w_i}\ .$$

Furthermore $\Omega_W(U)$ is generated by the image of $d_W(U)$. Hence

$$\xi(U)\colon \Omega_W(U) \to \mathcal{O}_W(U),\ \alpha_1 dw_1 + \ldots + \alpha_m dw_m \mapsto \alpha_i,$$

is as desired.

Before we treat the general case we prove a simple

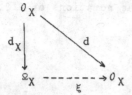

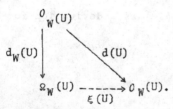

Lemma. Let $W \subset \mathbb{C}^m$ be an open set with coordinates $w_1,\dots,w_m$ and let $X \hookrightarrow W$ be a closed complex subspace defined by the ideal $I \subset \mathcal{O}_W$. Put $d = d_W$ and denote by

$$F := \mathcal{O}_W \cdot dI \subset \Omega_W$$

the $\mathcal{O}_W$-submodule generated by $d\,I$. Then $\mathcal{O}_W \cdot d\,I$ is $\mathcal{O}_W$-coherent and

$$I \cdot \Omega_W \subset \mathcal{O}_W \cdot dI \,.$$

Proof of the lemma. F_p is, by definition of F, the submodule of $\Omega_{W,p}$ generated by

$$\{df_p \colon f_p \in I_p\}$$

for every $p \in W$. The reader should be aware that in general F_p is bigger than the submodule of $\Omega_{W,p}$ generated by

$$\{df_{1,p},\dots,df_{k,p}\},$$

if $f_{1,p},\dots,f_{k,p}$ are generators of I_p.
For the convenience of the writer let us assume W is a polydisk. Then we may argue with global functions on W by means of Theorem B (0.37). Assume that $f_1,\dots,f_k \in \mathcal{O}_W(W)$ are generators of I. Then we claim that

$$df_\varkappa \text{ and } f_\varkappa dw_\mu \text{ for } \varkappa = 1,\dots,k \text{ and } \mu = 1,\dots,m$$

are generators of F over W.
The product rule for derivations yields

$$f_\varkappa \cdot dw_\mu = d(f_\varkappa \cdot w_\mu) - w_\mu \cdot df_\varkappa \in F(W).$$

If $f \in I(W)$, then

$$f = g_1 f_1 + \dots + g_k f_k \text{ with } g_1,\dots,g_k \in \mathcal{O}_W(W).$$

In order to prove that df is generated by our given forms we may suppose $f = g_\varkappa f_\varkappa$ for some $\varkappa$, since d is additive. Then

$$df = g \cdot df_\varkappa + f_\varkappa dg = g \cdot df_\varkappa + \frac{\partial g}{\partial w_1} f_\varkappa dw_1 + \dots + \frac{\partial g}{\partial w_m} f_\varkappa dw_m$$

and the coherence of F follows.
The second assertion follows immediately from

$$f_\varkappa(\omega_1 dw_1 + \dots + \omega_m dw_m) = \omega_1 f_\varkappa dw_1 + \dots + \omega_m f_\varkappa dw_m,$$

hence the lemma is proved.

For arbitrary X the question is local, hence we may assume $X \hookrightarrow W \subset \mathbb{C}^m$ is a closed subspace of an open set. If $I \subset \mathcal{O}_W$ denotes the defining ideal we consider the diagram

$$\begin{array}{ccccccccc} 0 & \longrightarrow & I & \longrightarrow & \mathcal{O}_W & \xrightarrow{\rho} & \mathcal{O}_X & \to & 0 \\ & & \downarrow & & \downarrow d_W & & \downarrow d_X & & \\ 0 & \to & \mathcal{O}_W \cdot d_W I & \hookrightarrow & \Omega_W & \xrightarrow{\sigma} & \Omega_X & \to & 0, \end{array} \qquad (*)$$

where ρ is the canonical homomorphism and σ, d_X are to be defined. For open $U \subset W$ and $V = X \cap U$ we assume $\omega \in \Omega_W(U)$ is given. Then we have a commutative diagram

$$\begin{array}{ccc} U \times \mathbb{C}^m & \xrightarrow{\omega} & U \times \mathbb{C} \\ \uparrow & & \uparrow \\ TV & \xrightarrow{\omega'} & V \times \mathbb{C}, \end{array}$$

where ω' is the restriction of ω, hence we may define

$$\sigma(U)\omega := \omega' \in \Omega_X(V).$$

From the definition of $TV \hookrightarrow U \times \mathbb{C}^m$ (see 2.5) it follows immediately that $\mathit{Ker}\ \sigma = \mathcal{O}_W \cdot dI$. Obviously σ is an epimorphism, hence the second line of (*) is exact.

With U and V as above assume $f \in \mathcal{O}_X(V)$ is given and $F \in \mathcal{O}_W(U)$ represents f. Then we define

$$d_X(V)f := \sigma(U)(d_W(U)F).$$

Since the right hand side is always zero if $F \in I(U)$, d_X is well defined in this way. It is easy to check that this is a derivation satisfying the universal property.

Corollary 1. For any complex space X there is a canonical isomorphism

$$T_X(X) \to \mathrm{Der}\ \mathcal{O}_X.$$

In other words, a vector field is the same thing as a derivation of the structure sheaf.

Proof. We have

$$\mathrm{Hom}(X \times \mathbb{C}, TX) = T_X(X) \quad \text{and}$$

$$\mathrm{Hom}(X \times \mathbb{C}, TX) = \mathrm{Hom}(\Omega_X, \mathcal{O}_X);$$

hence the assertion follows from the theorem.

Corollary 2. Let $\iota: Y \hookrightarrow X$ be a closed complex subspace defined by the ideal $I \subset \mathcal{O}_X$, and denote by $\mathcal{O}_X \cdot d_X I \subset \Omega_X$ the submodule generated by $d_X I$. Then

$$\Omega_X / \mathcal{O}_X \cdot d_X I$$

is an $\mathcal{O}_Y$-module and there is a canonical isomorphism

$$\Omega_X / \mathcal{O}_X \cdot d_X I \to \Omega_Y .$$

2.11. We shall now give conditions for a vector field to admit a restriction to a subspace or to the fibres of a holomorphic map.
Let us first assume we have an open subset $W \subset \mathbb{C}^m$. Then $TW = W \times \mathbb{C}^m$ and for a vector field $\Delta: W \to TW$ there are $\Delta_1, \dots, \Delta_m \in \mathcal{O}_W(W)$ such that

$$\Delta(w) = (w, \Delta_1(w), \dots, \Delta_m(w))$$

for all $w \in W$. By abuse of notation we sometimes write $\Delta = (\Delta_1, \dots, \Delta_m)$. Then the canonical isomorphism $\mathcal{T}_W \to \operatorname{Der} \mathcal{O}_W$ from 2.10 is given by

$$\mathcal{T}_W(U) \to \operatorname{Der} \mathcal{O}_W(U),\ U \subset W \text{ open},$$

$$\Delta \mapsto \Delta_1 \frac{\partial}{\partial w_1} + \dots + \Delta_m \frac{\partial}{\partial w_m} .$$

Now let $X \hookrightarrow W \subset \mathbb{C}^m$ be a closed complex subspace defined by the ideal $I \subset \mathcal{O}_W$ with generators $f_1, \dots, f_k \in \mathcal{O}_W(W)$ and assume a vector field

$$\Delta: W \to TW$$

is given. Since $TX \hookrightarrow TW = W \times \mathbb{C}^m$ is the closed complex subspace generated by

$$f_1, \dots, f_k \text{ and } \frac{\partial f_\varkappa}{\partial w_1} s_1 + \dots + \frac{\partial f_\varkappa}{\partial w_m} s_m \text{ for } \varkappa = 1, \dots, k,$$

(see 2.5) according to 0.17 the vector field Δ on W admits a restriction to a vector field

$$\delta: X \to TX$$

if and only if

$$\Delta_1 \frac{\partial f_\varkappa}{\partial w_1} + \dots + \Delta_m \frac{\partial f_\varkappa}{\partial w_m} \in I(W) \text{ for } \varkappa = 1, \dots, k. \qquad (*)$$

By using assertion a) of the remark below it is easy to check that this condition is equivalent to

$$\Delta(I) \subset I,$$

where $\Delta: \mathcal{O}_X \to \mathcal{O}_X$ also denotes the derivation canonically associated to the vector field Δ.

Remark. Let R be a $\mathbb{C}$-algebra, $\mathfrak{a} \subset R$ an ideal and $\delta: R \to R$ a derivation.

a) If $\mathfrak{a}$ is generated by $a_1,\dots,a_k$ and $\delta(a_\varkappa) \in \mathfrak{a}$ for $\varkappa = 1,\dots,k$, then $\delta(\mathfrak{a}) \subset \mathfrak{a}$.

b) δ induces a derivation $\delta': R/\mathfrak{a} \to R/\mathfrak{a}$ if and only if $\delta(\mathfrak{a}) \subset \mathfrak{a}$.

The proof is left to the reader.

By using condition (*) above or by part b) of the above remark and direct algebraic considerations we obtain the

Proposition. Let $\delta: X \to TX$ be a vector field on a complex space X and let $X' \hookrightarrow X$ be a closed complex subspace defined by the ideal $I \subset \mathcal{O}_X$. Then δ admits a restriction to a vector field

$$\delta': X' \to TX'$$

on X' if and only if $\delta(I) \subset I$, where

$$\delta: \mathcal{O}_X \to \mathcal{O}_X$$

also denotes the derivation canonically associated to the vector field δ (see 2.10).

Now we want to restrict a vector field not only to a single subspace but to the fibres of a holomorphic map. Let us first assume we have given a holomorphic map

$$\varphi: X \to \mathbb{C}^n,$$

where X is an arbitrary complex space, and a vector field

$$\delta: X \to TX.$$

According to 0.19 we have $\varphi = (\varphi_1,\dots,\varphi_n)$ with $\varphi_1,\dots,\varphi_n \in \mathcal{O}_X(X)$. Then

$$T(X/\mathbb{C}^n) \hookrightarrow TX$$

is the linear subspace over X generated by the Pfaffian forms

$$d\varphi_1,\dots,d\varphi_n.$$

Hence there is a factorization

$$X \to T(X/\mathbb{C}^n) \hookrightarrow TX$$

of δ if and only if

$$\delta\circ d\varphi_1 = \dots = \delta\circ d\varphi_n = 0. \qquad (**)$$

If we further assume $X \hookrightarrow W \subset \mathbb{C}^m$ is a closed subspace of some open subset, defined by the ideal $I \subset \mathcal{O}_W$, φ is generated by

$$\Phi = (\Phi_1,\ldots,\Phi_n): W \to \mathbb{C}^n$$

and δ is generated by

$$\Delta = (\Delta_1,\ldots,\Delta_m): W \to TW = W \times \mathbb{C}^m,$$

then condition (**) is equivalent to

$$\Delta_1 \frac{\partial\Phi_\nu}{\partial w_1} + \ldots + \Delta_m \frac{\partial\Phi_\nu}{\partial w_m} \in I(W) \text{ for } \nu = 1,\ldots,n.$$

The criterion just derived may be restated as follows:

Proposition. Assume we have given a holomorphic map $\varphi: X \to Y$, a vector field

$$\delta: X \to TX$$

and $p \in X$. Let $a_1,\ldots,a_n \in \mathcal{O}_{Y,\varphi(p)}$ be generators of $\mathfrak{m}_{Y,\varphi(p)}$, denote by

$$\tilde{\varphi}_p: \mathcal{O}_{Y,\varphi(p)} \to \mathcal{O}_{X,p}$$

the canonical homomorphism and by

$$\delta_p: \mathcal{O}_{X,p} \to \mathcal{O}_{X,p}$$

the germ of the derivation corresponding to δ. Then the following conditions are equivalent:

i) There is an open neighbourhood U of p in X and a factorization

$$U \to T(U/Y) \hookrightarrow TU$$

of $\delta|U$.

ii) $\delta_p(\tilde{\varphi}_p a_\nu) = 0$ for $\nu = 1,\ldots,n$.

2.12. Several "regularity criterions" will be an easy consequence of the following

Theorem. If X is a complex space then for $p \in X$ the following conditions are equivalent:

i) There is a local analytic algebra R and an isomorphism

$$\mathcal{O}_{X,p} \to \mathbb{C}[\langle z_1,\ldots,z_k\rangle] \hat{\otimes} R.$$

ii) There is an open neighbourhood U of p in X, a complex space X', an open subset $Z \subset \mathbb{C}^k$ and a biholomorphic map

$$U \to Z \times X'.$$

iii) There are germs of vector field $\xi_1,\ldots,\xi_k \in T_{X,p}$ such that

$$\xi_1(p),\ldots,\xi_k(p) \in T_pX$$

are linearly independent.

In case X is reduced this was proved by ROSSI [121]. Our proof will make use of the

Remark. For $m \geq 1$ take $f \in \mathbb{C}[\langle z,w_2,\ldots,w_m\rangle]$ and expand it as

$$f = f_o + f_1 z + f_2 z^2 + \ldots \text{ with } f_\mu \in \mathbb{C}[\langle w_2,\ldots,w_m\rangle].$$

Then

$$f_o = \sum_{\nu=0}^{\infty} \frac{(-1)^\nu}{\nu!} \cdot \frac{\partial^\nu f}{\partial z^\nu} \cdot z^\nu .$$

Such a formula has often been used (see e.g. [156], p. 526).

Proof of the remark. It is sufficient to show that the equality holds for formel power series. This follows from

$$\sum_{\nu=0}^{\infty} \frac{(-1)^\nu}{\nu!} \cdot \frac{\partial^\nu f}{\partial z^\nu} \cdot z^\nu$$

$$= \sum_{\nu=0}^{\infty} \frac{(-1)^\nu}{\nu!} \sum_{\mu=\nu}^{\infty} \frac{\mu!}{(\mu-\nu)!} f_\mu z^\mu$$

$$= \sum_{\mu=0}^{\infty} \sum_{\nu=0}^{\mu} (-1)^\nu \binom{\mu}{\nu} f_\mu z^\mu = f_o .$$

For the last equality we have used

$$\sum_{\nu=0}^{\mu} (-1)^\nu \binom{\mu}{\nu} = 0 \text{ for } \mu \geq 1.$$

Proof of the theorem. The equivalence of i) and ii) is a direct consequence of 0.21 and 0.29. The implication ii) ⇒ iii) follows from 2.8.

In order to prove iii) ⇒ ii) we first consider the case $k = 1$. We may assume that X is a closed subspace of some open subset

$$W = Z \times W' \subset \mathbb{C}^m \text{ with coordinates } (z,w_2,\ldots,w_m)$$

and that ξ is represented by $\delta \in \mathcal{T}_X(X)$ which is induced by the vector field $\Delta \in \mathcal{T}_W(W)$. After a change of coordinates in W we may obtain (see e.g. [24] p. 89)

$$\Delta = \frac{\partial}{\partial z} .$$

Now we define $X' = W' \cap X$ (see 0.27), i.e. the closed subspace of X defined by the residue class of the coordinate function z.
We want to prove $X = Z \times X'$ (after shrinking X sufficiently). Denote by

$I \subset \mathcal{O}_{W,p}$ the ideal of $X \hookrightarrow W$

$I' \subset \mathcal{O}_{W',p}$ the ideal of $X' \hookrightarrow W'$

$I^* = (I' \otimes 1)\cdot\mathcal{O}_{W,p} \subset \mathcal{O}_{W,p}$ the ideal of $X' \times Z \hookrightarrow W$.

Then we have to show $I = I^*$.
Take $f \in I$ and expand it as

$$f = f_0 + f_1 z + f_2 z^2 + \ldots \text{ with } f_\mu \in \mathbb{C}[\langle w_2,\ldots,w_m\rangle].$$

Since Δ admits a restriction to a vector field δ on X, proposition 2.11 implies

$$\frac{\partial^\mu f}{\partial z^\mu} \in I \text{ for } \mu = 0,1,2,\ldots . \qquad (*)$$

On the other hand

$$f_\mu = \frac{1}{\mu!}\left.\frac{\partial^\mu f}{\partial z^\mu}\right|_{z=0}, \text{ hence } f_\mu \in I'.$$

Since $I^* \subset \mathcal{O}_{W,p}$ is closed (see [19], [50]), we obtain $f \in I^*$.

In order to prove $I^* \subset I$ we show $I' \otimes 1 \subset I$. Assume $f_0 \in I'$. Then there exists some

$$f = f_0 + f_1 z + f_2 z^2 + \ldots \in I.$$

This implies (*) as above and our Remark yields $f_0 \in I$.
For arbitrary k we may repeat this procedure. With the notations as above assume we have additional germs $\xi_2,\ldots,\xi_k$. By omitting the component with respect to the coordinate z we obtain germs $\xi_2',\ldots,\xi_k' \in T_{X',p}$ such that

$$\xi_2'(p),\ldots,\xi_k'(p) \in T_pX'$$

are linearly independent and the induction works.

2.13. We are going to prove the analyticity of several kinds of "singular loci" for complex spaces and holomorphic maps. This will be a consequence of the following considerations on coherent sheaves of modules.

Assume we have given a complex space X, a coherent $\mathcal{O}_X$-module $\mathcal{F}$ and $p \in X$. Then

$$\mathcal{F}(p) := \mathcal{F}_p/\mathfrak{m}_{X,p}\mathcal{F}_p = \mathcal{F}_p \otimes_{\mathcal{O}_{X,p}} \mathbb{C}$$

is a finite dimensional $\mathbb{C}$-vector space. We call

$$\operatorname{cork}_p \mathcal{F} := \dim_{\mathbb{C}} \mathcal{F}(p)$$

the *corank* of $\mathcal{F}$ at p. Then for any $r \in \mathbb{N}$ there is a certain *singular locus*

$$A_r(\mathcal{F}) := \{p \in X : \operatorname{cork}_p \mathcal{F} > r\}.$$

Lemma. For any coherent $\mathcal{O}_X$-module $\mathcal{F}$ and any $r \in \mathbb{N}$ the subset $A_r(\mathcal{F}) \subset X$ is analytic.

Proof. The question being local we may assume we have an exact sequence

$$\mathcal{O}_X^l \xrightarrow{\alpha} \mathcal{O}_X^k \longrightarrow \mathcal{F} \longrightarrow 0$$

which yields an exact sequence

$$\mathbb{C}^l \xrightarrow{\alpha(p)} \mathbb{C}^k \longrightarrow \mathcal{F}(p) \longrightarrow 0$$

for every point $p \in X$ and consequently

$$A_r(\mathcal{F}) = \{p \in X : \operatorname{rank} \alpha(p) < k-r\}.$$

Since α is given by a holomorphic $(k\times l)$-matrix on X, the subset $A_r(\mathcal{F}) \subset X$ is the common set of zeroes of all the subdeterminants of order k-r.

For different concepts of rank we refer to [126], [50].

Corollary. Let X be a reduced complex space, $\mathcal{F}$ a coherent $\mathcal{O}_X$-module. Define

$$c(\mathcal{F}) := \min\{\operatorname{cork}_p \mathcal{F} : p \in X\} \text{ and}$$
$$A(\mathcal{F}) := \{p \in X : \operatorname{cork}_p \mathcal{F} > c(\mathcal{F})\}.$$

Then $A(\mathcal{F}) \subset X$ is a proper analytic subset and $\mathcal{F}$ is locally free on $X \setminus A(\mathcal{F})$.

If, in addition, X is irreducible then

$$A(\mathcal{F}) = \{p \in X : \mathcal{F} \text{ is not free at } p\}.$$

Proof. Outside A(F) the rank of F is constant and equal to c(F). The lemma of 1.8. implies that F is locally free outside A(F).
If X is irreducible and $p \in A(F)$ then every neighbourhood of p in X contains points not in A(F). Hence F cannot be free at p.

2.14. For an arbitrary coherent module F over a complex space X we define another *singular locus*

$$S(F) = \{p \in X: F \text{ is not free at } p\}.$$

By 0.5 and 0.9 F is free at p if and only if F_p is a free $\mathcal{O}_{X,p}$-module.

Proposition. If X is a complex space and F is a coherent $\mathcal{O}_X$-module, then $S(F) \subset X$ is an analytic set.

First proof. The question being local we may assume we have an exact sequence

$$\mathcal{O}_X^{\ell} \to \mathcal{O}_X^{k} \overset{\alpha}{\to} F \to 0.$$

For an arbitrary homomorphism $\beta: \mathcal{O}_X^r \to F$ we define the set

$$S(\beta) := \mathrm{supp}(\mathit{Ker}\ \beta) \cup \mathrm{supp}(\mathit{Coker}\ \beta) \subset X$$

which is analytic in X by 0.13. Now we claim

$$S(F) = \bigcap_{r,\beta} S(\beta)$$

where r and β are arbitrary. Since the intersection of an arbitrary number of analytic sets is analytic, this proves the assertion.
If $p \notin S(\beta)$ for some β then $\beta_p: \mathcal{O}_{X,p}^r \to F_p$ is an isomorphism and $p \notin S(F)$.
If $p \notin S(F)$ then F_p is free. Put $r := \mathrm{cork}_p F$. Then by using α from above and the lemma of NAKAYAMA we may find

$$\beta: \mathcal{O}_X^r \to F \quad \text{such that } \beta_p: \mathcal{O}_{X,p}^r \to F_p$$

is surjective. By again using NAKAYAMAs lemma it is easy to see that β_p is an isomorphism (compare the proof of Proposition 3.12). Hence $p \notin S(\beta)$.

Second proof (due to H.W.Schuster). We may assume we have an epimorphism

$$\alpha: \mathcal{O}_X^k \to F.$$

It induces a homomorphism of coherent $\mathcal{O}_X$-modules

$$\phi: \mathit{Hom}_{\mathcal{O}_X}(F,\mathcal{O}_X) \to \mathit{Hom}_{\mathcal{O}_X}(\mathcal{O}_X^k,\mathcal{O}_X).$$

Now it is quite elementary to prove $S(F) = \mathrm{supp}(\mathit{Coker}\ \phi)$, since F_p is free if and only if there is a section

$$\sigma: F_p \to \mathcal{O}_{X,p}^k \quad \text{of} \quad \alpha_p: \mathcal{O}_{X,p}^k \to F_p .$$

<u>Corollary.</u> If L is a linear fibre space over a complex space X then

$\{p \in X$: L is not a vector bundle at $p\} \subset X$

is an analytic subset.

<u>2.15.</u> Finally we are able to prove a "differential regularity criterion" for complex spaces.

<u>Theorem.</u> For a point p of a complex space X the following conditions are equivalent:

i) X is a manifold at p.

ii) The tangent space TX over X is a vector bundle at p.

iii) The sheaf Ω_X of Pfaffian forms is free at p.

<u>Corollary.</u> For any complex space X the *singular locus*

$S(X) = \{p \in X$: X is not a manifold at $p\} \subset X$

is an analytic subset.

<u>Proof.</u> The corollary follows immediately by using 2.13.

In the theorem everything except ii) $\Rightarrow$ i) is clear. Denote by n the rank of TX at p, i.e.

$$n = \dim_{\mathbb{C}} T_p X.$$

Theorem 2.12 yields a neighbourhood U of p in X, a closed complex subspace $X' \hookrightarrow U$ and an open subset $Z \subset \mathbb{C}^n$ such that

$$U = Z \times X'.$$

By proposition 2.3 we have

$$\operatorname{emdim}_p X = n, \text{ hence } \operatorname{emdim}_p X' = 0.$$

This implies that X' is the simple point $(\{p\},\mathbb{C})$ and consequently X is a manifold at p.

It should be observed that this theorem is false in general if the characteristic of the ground field is different from zero. Then in conditions ii) and iii) one has to assume in addition that the rank of TX or the corank of Ω_X at p are equal to the dimension of X at p (see [52], Exp. 14).

For the concept of *ordinary points* of a complex space and the proof that the set of not ordinary points is analytic (by using 2.12) we refer to [126].

<u>2.16.</u> <u>Proposition.</u> For a complex space X the following conditions are equivalent:

i) X is reduced.

ii) The singular locus $S(X) \subset X$ is analytically rare.

Proof. i) ⇒ ii). Using the decomposition of a reduced complex space into irreducible components (see e.g. [ ], [ ]) it is sufficient to prove that $S(X) \subset X$ is a proper analytic subset in case X is irreducible. But this is a direct consequence of the corollary in 2.13 and 2.15.

ii) ⇒ i). If X is not reduced then there is an open $U \subset X$, a function $0 \neq f \in \mathcal{O}_X(U)$ and a $k \in \mathbb{N}$ such that $f^k = 0$. Obviously

$$f|(U \setminus S(X)) = 0,$$

and consequently S(X) is not analytically rare.

2.17. Following the pattern of differential topology we use the Jacobian map in order to give a regularity criterion for holomorphic maps. First we introduce some useful notions.

Recall that for a holomorphic map $\varphi: X \to Y$ we denote by $T\varphi: TX \to \varphi^*TY$ its Jacobian map and by $T(X/Y) = \operatorname{Ker} T\varphi$ the tangent space in the fibre direction (2.6, 2.7).

For $p \in X$ we define the *corank* and *rank* of φ at p by

$$\operatorname{cork}_p\varphi := \dim_{\mathbb{C}} T_p(X/Y) = \dim_{\mathbb{C}} T_p X_{\varphi(p)} \quad \text{and}$$

$$\operatorname{rk}_p\varphi := \dim_{\mathbb{C}} T_p X - \dim_{\mathbb{C}} T_p(X/Y).$$

The lemma of 2.13 immediately implies

Lemma. For any holomorphic map $\varphi: X \to Y$ and $r \in \mathbb{N}$ the set

$$\operatorname{Sing}^k(\varphi) := \{p \in X : \operatorname{cork}_p\varphi > k\}$$

is analytic.

Of course this is not true in general for the sets

$$\{p \in X : \operatorname{rk}_p\varphi < k\}$$

unless X is a manifold. In this case there exists a useful estimate for the dimension (see [15], [135]).

Proposition. Let $\varphi: X \to Y$ be a holomorphic map between complex manifolds and assume $\dim X \geq \dim Y$. Put

$$S := \{p \in X : \operatorname{rk}_p\varphi < \dim Y\} \subset X .$$

Then for any $p \in S$ we have $\dim_p S \geq \dim_{\varphi(p)} Y - 1$.

In the proof we use some elementary facts about the dimension of analytic sets (see 3.1 for basic facts).

Remark 1. Consider an open subset $U \subset \mathbb{C}^n$ and analytic sets $A,B \subset U$. Take $p \in A \cap B$ and assume B is non-singular in p. Then

$$\operatorname{codim}_p(A;U) \geq \operatorname{codim}_p(A \cap B;B) .$$

Proof. We may assume $p = 0$ and B is a linear subspace. Then B contains a linear subspace C of dimension $\operatorname{codim}_0(A \cap B;B)$ such that 0 is an isolated point of $A \cap B \cap C$. Hence 0 is an isolated point of $A \cap C$ and this implies the assertion.

Remark 2. Let $\varphi: X \to Y$ be a holomorphic map between complex manifolds. If $A \subset Y$ is an analytic subset then for any $p \in \varphi^{-1}(A)$ we have

$$\operatorname{codim}_{\varphi(p)}(A;Y) \geq \operatorname{codim}_p(\varphi^{-1}(A);X) .$$

Proof. The subset

$$\Gamma_\varphi := \{(x,y) \in X \times Y : \; y = \varphi(x)\} \subset X \times Y$$

is non-singular since the restriction

$$\pi: \Gamma_\varphi \to X$$

of the projection is biholomorphic. Obviously

$$\varphi^{-1}(B) = \pi(\Gamma_\varphi \cap (B \times X)) .$$

Hence the assertion follows from Remark 1.

Remark 3. Denote by $M(n{\times}m;\mathbb{C})$ the $m{\cdot}n$-dimensional vector space of $(n{\times}m)$-matrices and put

$$E := \{A \in M(n{\times}m;\mathbb{C}) : \text{rank } A < n\} .$$

If $n \leq m$, then $E \subset \mathbb{C}^{mn}$ is an irreducible analytic set of codimension $m - n + 1$.

Proof. That the codimension of E is in fact equal to $m - n + 1$ may easily be seen by "counting constants": If rank $A < n$ then only the first $n - 1$ rows and columns may be selected arbitrarily. The remaining $m - n + 1$ entries must be linear combinations.

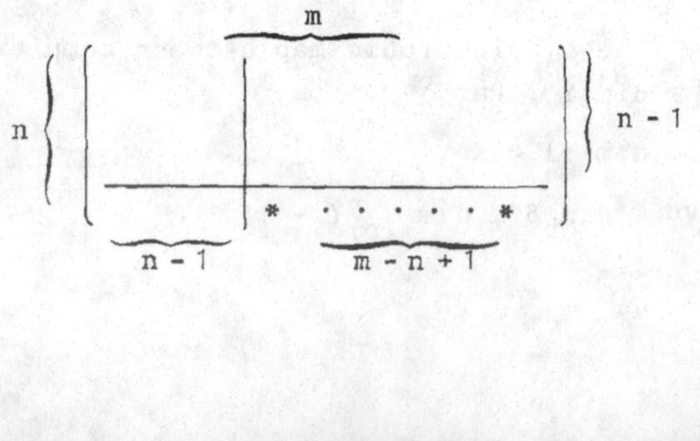

In order to prove the irreducibility put

$$E^k_{n,m} := \{A \in M(n\times m;\mathbb{C}) : \text{rank } A \le k\} .$$

It is elementary to construct a holomorphic map

$$A := E^{n-1}_{n,m} \setminus E^{n-2}_{n,m} \longrightarrow \mathbb{P}_{n-1}(\mathbb{C})$$

which is a fibre bundle with typical fibre $M(m\times(n-1);\mathbb{C}) \setminus E^{n-2}_{m,n-1}$.
Hence A is a complex manifold with

$$\dim A = n-1 + m(n-1) = n\cdot m - (m-n+1) .$$

Since E is the closure of A the assertion follows.

<u>Proof of the proposition.</u> We use the Jacobian map

$$T\varphi: TX \longrightarrow \varphi^*TY .$$

The question being local we may use local coordinates and assume $T\varphi$ induces a holomorphic map

$$\Phi: X \longrightarrow M(n\times m;\mathbb{C})$$

(where $m = \dim X$ and $n = \dim Y$) such that

$$S = \Phi^{-1}(E) .$$

Hence Remarks 2 and 3 imply the assertion.

It is much more difficult to derive estimates for the dimension of the set $\varphi(S)$ of "critical values" of (see e.g. [123]).

<u>2.18.</u> The locally most "regular" complex spaces are complex manifolds. For holomorphic maps a concept of local "regularity" is defined as follows.
A holomorphic map $\varphi: X \to Y$ between arbitrary complex spaces is called a k-*mersion at* $p \in X$, if one can find the following items: open neighbourhoods $U \subset X$ of p, $V \subset Y$ of $\varphi(p)$ with $\varphi(U) \subset V$, an open subset $Z \subset \mathbb{C}^k$, a closed complex subspace $V' \hookrightarrow V$ and a biholomorphic map $\psi: U \to Z \times V'$ such that the diagram

$$\begin{array}{ccc} U & \xrightarrow{\psi} & Z \times V' \\ {\scriptstyle\varphi}\downarrow & & \downarrow{\scriptstyle\pi_2} \\ V & \hookleftarrow & V' \end{array}$$

commutes.

φ is called a k-*mersion* if it is a k-mersion at every point, and a *mersion* if at every point it is a k-mersion for some k.
If we may choose $V' = V$, φ is called a *submersion* at p. Obviously a 0-mersion is just an immersion.

As one immediately sees the fibres of a mersion are always complex manifolds. Since by 2.7

$$T_p X_{\varphi(p)} = T_p(X/Y),$$

a k-mersion has corank k.
The reader should be aware of the fact that the composition of mersions need not be a mersion.

Example. Take an immersion and a submersion

$$\varphi: \mathbb{C} \to \mathbb{C}^2,\ z \mapsto (z^2, z), \text{ and } \psi: \mathbb{C}^2 \to \mathbb{C},\ (x,y) \mapsto x.$$

Then $\psi \circ \varphi$ is not a mersion at 0.

2.19. As in the absolute case (2.15), there is also a relative version of a differential regularity criterion.

Theorem. Assume we have a holomorphic map $\varphi: X \to Y$ and a point $p \in X$. Then the following conditions are equivalent:
 i) φ is a mersion at p.
ii) $T(X/Y)$ is a vector bundle at p.

Proof. i) $\Rightarrow$ ii) follows immediately from the definition of a mersion.
ii) $\Rightarrow$ i). By definition $T(X/Y) \hookrightarrow TX$ is a linear subspace. If it has rank k at p we may apply 2.15; the problem being local we may assume we have the following diagram:

$$\begin{array}{ccccc} Z \times X' = X & \hookrightarrow & W = Z \times W' & \subset & \mathbb{C}^k \times \mathbb{C}^{m-k} \\ {\scriptstyle\pi}\downarrow \quad\ \downarrow{\scriptstyle\varphi} & & \downarrow{\scriptstyle\Phi} & & \\ X' \xrightarrow{\ \varphi'\ } Y & \hookrightarrow & V & \subset & \mathbb{C}^n. \end{array}$$

Here X resp. Y are closed complex subspaces of open subsets $W \subset \mathbb{C}^m$ with p = 0 resp. $V \subset \mathbb{C}^n$ and $\Phi = (\Phi_1,\ldots,\Phi_n)$ generates φ (see 0.22). If

$(z_1,\ldots,z_k,w_{k+1},\ldots,w_m)$ are coordinates in $Z \times W'$

we may assume (as in the proof of 2.15) that the vector fields

$$\frac{\partial}{\partial z_1}, \ldots, \frac{\partial}{\partial z_k} \text{ on } W$$

induce vector fields $\delta_1,\ldots,\delta_k\colon X \to T(X/Y)$ such that

$\delta_1(p),\ldots,\delta_k(p)$ is a basis of $T_p(X/Y)$.

If $X' = X \cap (\{0\} \times W')$ is defined as the analytic intersection (see 0.27), i.e. the closed complex subspace of X generated by the residue classes of $z_1,\ldots,z_k$, we denote by π the canonical projection. What we finally have to prove is the existence of a holomorphic map

$\varphi'\colon X' \to Y$

which is an immersion at p such that $\varphi = \varphi' \circ \pi$.

We define $\varphi' := \varphi|X'$. In order to prove $\varphi = \varphi' \circ \pi$, we show that we can choose the functions $\Phi_1,\ldots,\Phi_n$ independent of $z_1,\ldots,z_k$. Denote by

$$\psi_1,\ldots,\psi_n \in \mathcal{O}_{W,o} = \mathbb{C}[\langle z_1,\ldots,z_k,w_{k+1},\ldots,w_m\rangle]$$

the germs of $\Phi_1,\ldots,\Phi_n$ at p = 0. For i = 1,...,n there is an expansion

$$\psi_i = \sum_{r \in \mathbb{N}^k} \psi_i^{(r)} z^r \text{ with } \psi_i^{(r)} \in \mathbb{C}[\langle w_{k+1},\ldots,w_m\rangle].$$

Now we apply what we have proved in 2.11. If $I \subset \mathcal{O}_{W,o}$ denotes the ideal of X at o, then

$$\frac{\partial \psi_i}{\partial z_j} \in I \text{ for } i = 1,\ldots,n \text{ and } j = 1,\ldots,k,$$

since the vector fields $\frac{\partial}{\partial z_j}$ may be restricted to vector fields along the fibres of φ. Again by applying 2.11 we obtain

$$\frac{\partial^{|r|}\psi_i}{\partial z^r} \in I \text{ for all } r \in \mathbb{N}^k \text{ with } |r| \geq 1$$

If $I' \subset \mathbb{C}[\langle w_{k+1},\ldots,w_m\rangle]$ denotes the ideal of X', then

$$r!\psi_i^{(r)} = \left.\frac{\partial^{|r|}\psi_i}{\partial z^r}\right|_{z=0} \in I' \text{ for } |r| \geq 1.$$

Since $X = Z \times X'$ this implies $\psi_i^{|r|} \in I$ for $|r| \geq 1$. $I \subset \mathcal{O}_{W,o}$ being closed this implies

$$\psi_i - \psi_i^{(o)} \in I \text{ for } i = 1,\ldots,n.$$

Hence we may shrink W sufficiently and replace $\Phi_1,\ldots,\Phi_n$ by representatives $\Phi_1^{(o)},\ldots,\Phi_n^{(o)} \in \mathcal{O}_W(W)$ of $\psi_1^{(o)},\ldots,\psi_n^{(o)}$ without changing the holomorphic map induced on X.
It remains to show that φ' is an immersion at p. As we have seen $\operatorname{Ker} T\varphi = \operatorname{Ker} T\pi_2$. Hence $T_p\varphi'$ is injective and the assertion follows from 2.4.

<u>Corollary 1.</u> If $\phi: X \to Y$ is a holomorphic map between complex spaces then the subset

$\{p \in X: \phi \text{ is not a mersion at } p\} \subset X$

is analytic.

This follows immediately by using 2.14.

<u>Corollary 2.</u> Let $\phi: X \to Y$ be a holomorphic map and assume X is reduced. Then for $p \in X$ the following conditions are equivalent:

i) ϕ is a mersion at p.

ii) There is a neighbourhood U of p in X such that $\operatorname{cork}_x\phi$ is constant for all $x \in U$.

By 1.8, condition ii) expresses that $T(X/Y)$ is a vector bundle at p (see also [70]).
For more regularity criterions we refer to [30], [124], [125].

<u>2.20.</u> Let $\iota: Y \to X$ be a closed complex subspace. Then a holomorphic map $\rho: X \to Y$ is called a *retraction* if $\rho \circ \iota = \mathrm{id}_Y$, i.e. the diagram

$$\begin{array}{ccc} Y & \xrightarrow{\iota} & X \\ & \searrow^{\mathrm{id}_Y} & \downarrow \rho \\ & & Y \end{array}$$

commutes.

As an application of our differential tools it is easy to prove (see [121])

<u>Proposition.</u> If $\iota: Y \hookrightarrow X$ is a closed complex subspace of a complex manifold and if there exists a holomorphic retraction

$\rho: X \to Y$

then Y is a complex manifold and ρ is a submersion.

Proof. The problem being local we may assume that for a given point $p \in Y$ there are vector fields

$\delta_1, \ldots, \delta_k: X \to TX$

such that $\delta_1(p), \ldots, \delta_k(p) \in T_pX$ are a basis of the vector subspace $T_pY \subset T_pX$. Now the Jacobian map

$T\rho: TX \to \rho^*TY$

induces a homomorphism

$\iota^*T\rho: \iota^*TX \to \iota^*\rho^*TY = (\rho \circ \iota)^*TY = TY$

of linear spaces over Y which is a retraction of

$TY \hookrightarrow \iota^*TX$.

Obviously any vector field

$\delta: X \to TX$

induces a holomorphic section

$\delta': Y \to \iota^*TX$

and we obtain a vector field

$\delta'' := (\iota^*T\rho) \circ \delta': Y \to TY$

on Y such that

$\delta''(p) = \delta(p)$ in case $\delta(p) \in T_pY$.

Consequently the values at p of the vector fields

$\delta_1'', \ldots, \delta_k'': Y \to TY$

are a basis of T_pY and 2.12 implies that Y is a manifold at p. As we have just seen for $p \in Y$ the map

$T_p\rho: T_pX \to T_pY$

is surjective. Hence $\mathrm{cork}_p\rho = m - k$, where $m := \dim_p X$. The semicontinuity of the corank (2.17) implies

$\mathrm{cork}_x\rho \leq m - k$ for $x \in X$

if X is sufficiently small. On the other hand we know from linear algebra

$\mathrm{cork}_x\rho \geq m - k$.

Hence close to p the corank of ρ is constant and Corollary 2 of 2.19 implies that ρ is a submersion at p.

It should be noted here that any complex subspace of a complex space admits a continuous neighbourhood-retraction. This is a consequence of the triangulation theorem for analytic sets (see [38], [100]). But a singular analytic subset of an open subset in $\mathbb{C}^n$ does not even admit a C^1-retraction (see [39]).

Of course our proposition can be proved with much less differential calculus. It is an easy consequence of the

Lemma. Let R and S be local analytic $\mathbb{C}$-algebras with maximal ideals $m \subset R$ and $n \subset S$, and assume there is an epimorphism

$$\phi: R \to S \quad \text{with a section} \quad \sigma: S \to R .$$

Then S is regular if R was regular.

Proof. If

$$R = \mathbb{C}[\langle x_1,\dots,x_k\rangle]$$

it is easy to see that we may assume

$$k = \dim_{\mathbb{C}} m/m^2 .$$

Obviously σ is injective. Since the derivative of σ

$$n/n^2 \to m/m^2$$

is surjective, σ is surjective (see e.g. [50], II.§3).

2.21. As another useful differential device we are going to define a *normal space* of a holomorphic map $\phi: X \to Y$. In 2.6 we have introduced the Jacobian map

$$T\phi: TX \to \phi^*TY$$

with kernel $T(X/Y) \hookrightarrow TX$, the tangent space of ϕ. By

$$N(X/Y) \hookrightarrow \phi^*TY$$

we denote the linear space over X which is the image of $T\phi$; we call it the *normal space of* ϕ. Then we have an exact sequence

$$0 \to T(X/Y) \to TX \to N(X/Y) \to 0 \qquad (*)$$

of linear spaces over X. This definition deserves some comment.

Let us first consider the example $X = Y = \mathbb{C}$ and

$$\phi: X \to Y , \quad x \mapsto x^2 .$$

Then we have

$$T\phi: X \times \mathbb{C} \to X \times \mathbb{C} \ , \quad (x,s) \mapsto (x,2xs) \ .$$

This map is not surjective but an epimorphism of linear fibre spaces. Hence in this case $N(X/Y) = X \times \mathbb{C}$.

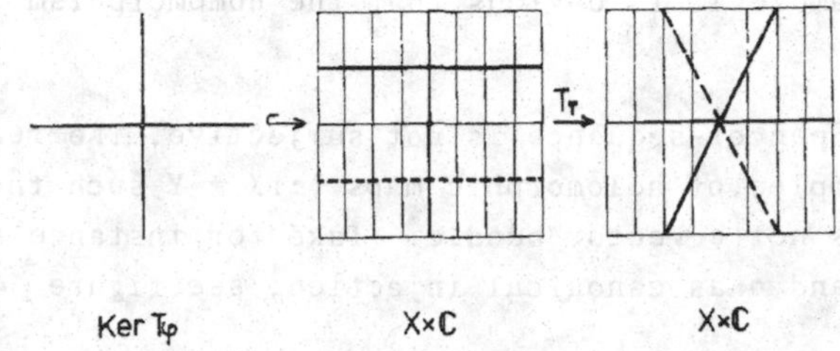

Figure 3

This is the same as the normal space of the identic map

$$\mathbb{C} \to \mathbb{C} \ , \quad x \mapsto x \ .$$

But a characteristic difference becomes apparent if we look at vector fields.
Let us first consider the exact sequence of coherent $\mathcal{O}_X$-modules which is dual to (*). If

$$D\phi: \ \psi^* \Omega_Y \to \Omega_X$$

denotes the dual of the Jacobian map (see 2.9), we denote by

$$\Omega_Y^* \subset \Omega_X$$

the image of $D\phi$ and by

$$\Omega_{X/Y} := \Omega_X / \Omega_Y^*$$

the sheaf of *Pfaffian forms along the fibres of* ϕ.
Then obviously the exact sequence

$$0 \to \Omega_Y^* \to \Omega_X \to \Omega_{X/Y} \to 0 \qquad (**)$$

is obtained from (*) by applying the functor

$$Hom_X(\ \ , X \times \mathbb{C})$$

(see 1.6). If we want to pass from (*) to vector fields we have to apply

$$Hom_X(X \times \mathbb{C}, \ \ ) \ .$$

This is the same as applying

$$Hom_{\mathcal{O}_X}(\ \ , \mathcal{O}_X)$$

to (**). As result we obtain the exact sequence

$$0 \to T_{X/Y} \to T_X \to N_{X/Y} ,$$

where $T_{X/Y}$ resp. $N_{X/Y}$ is the sheaf of holomorphic sections in T(X/Y) resp. N(X/Y). We call this sequence the *Kodaira-Spencer-sequence* of ϕ.

In our above example it is obvious that the homomorphism

$$T_X \to N_{X/Y}$$

of the Kodaira-Spencer-sequence is not surjective. The reader will easily find examples of holomorphic maps ϕ: X → Y such that the normal bundle N(X/Y) is not a vector bundle. (Take for instance a double point X, Y = ℂ and ϕ as canonical injection; see figure 4).

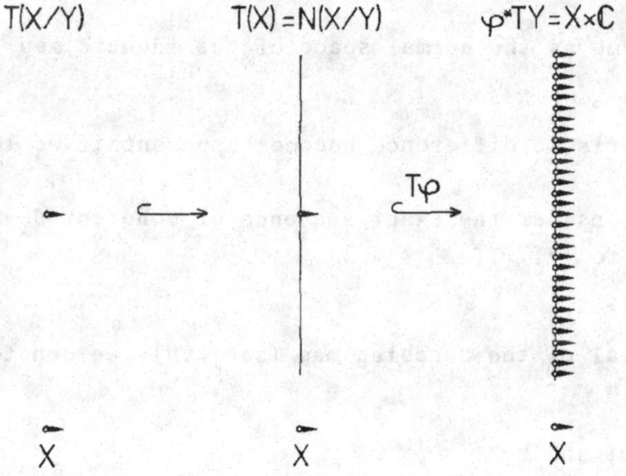

Figure 4

2.22. A holomorphic map ϕ: X → Y between arbitrary complex spaces is called *trivial at* p ∈ X if there is a neighbourhood U of p in X, V of ϕ(p) in Y with $\phi(U) \subset V$, a complex space X' and a biholomorphic map ψ such that the diagram

$$\begin{array}{ccc} U & \xrightarrow{\ \psi\ } & V \times X' \\ & {\scriptstyle \phi|U} \searrow & \downarrow {\scriptstyle \pi_1} \\ & & V \end{array}$$

is commutative. Here π_1 denotes the canonical projection.
ϕ is called *locally trivial*, if it is trivial at every point of X. In case Y is a manifold there is an important differential criterion for local triviality (see [45], [141]); this criterion ist in some sense transversal to 2.19.

Proposition. Let $\phi: X \to Y$ be a holomorphic map of the complex space X to the complex manifold Y. Then for $p \in X$ the following conditions are equivalent:

i) ϕ is trivial at p .

ii) The normal space N(X/Y) is trivial at p of rank $n := \dim_{\phi(p)} Y$ and the homomorphism

$$T_X \to N_{X/Y}$$

of the Kodaira-Spencer-sequence is surjective at p.

Proof. i) $\Rightarrow$ ii) is an immediate consequence of 2.8.
ii) $\Rightarrow$ i). We first consider the case n = 1. The question being local we may assume we have given a holomorphic function

$$\phi: X \to \mathbb{C} = Y ,$$

$X \hookrightarrow W \subset \mathbb{C}^m$ is a closed subspace of some open subset W, and ϕ is generated by

$$\Phi: W \to \mathbb{C} .$$

Now ii) implies that there is a vector field

$$\delta: X \to TX$$

whose residue class generates $N_{X/Y}$ at p ; then $\delta(p) \neq 0 \in T_pX$. We may further assume, δ is induced by a vector field

$$\Delta: W \to TW = W \times \mathbb{C}^m .$$

Since $\Delta(p) \neq 0 \in T_pW$ we may choose coordinates $z, w_2, \ldots, w_m$ in W such that (see [24], p. 89)

$$\Delta = \frac{\partial}{\partial z} .$$

According to 2.12 there is a splitting

$$X = Z \times X' , \qquad (\times)$$

where $Z \subset \mathbb{C}$ is open and $X' \hookrightarrow W' \subset \mathbb{C}^{m-1}$ (with coordinates $w_2, \ldots, w_m$) is a closed complex subspace.
By assumption the sequence

$$0 \to T_{X/Y,p} \to T_{X,p} \to N_{X/Y,p} \to 0$$

is exact. Since N(X/Y) is trivial at p this implies that the sequence

$$0 \to T_p(X/Y) \to T_pX \to N_p(X/Y) \to 0$$

of $\mathbb{C}$-vector spaces is exact (just consider the values at p of the vector fields!).
Since our given vector field δ is not in the fibre direction we have

$\delta(p) \notin T_p(X/Y)$. According to 2.11 this means

$$0 \neq (\delta\phi)(p) = (\Delta\Phi)(p) = \frac{\partial\Phi}{\partial z}(p) \in \mathbb{C} .$$

Consequently $(\Phi, w_2, \ldots, w_m)$ is again a coordinate system in W and obviously the splitting (×) is preserved. This proves i) in case $n = 1$.

For the general case it is sufficient to iterate this procedure. We leave the simple details to the reader.

It is easy to show that $N(X/Y)$ is trivial in case φ is flat and reduced (see [141]).

Appendix. Normal and Maximal Complex Spaces

In this section we give a brief outline of basic facts. For some complicated technical details we shall refer to the literature.

2.23. We first state the classical theorems concerning the extension of holomorphic functions into analytic subsets of complex manifolds. Recall that an analytic subset A of a reduced complex space X is rare if and only if locally its intersection with every irreducible component of X is a proper subset (see 0.43), i.e. $\operatorname{codim}_x A \geq 1$ for every $x \in X$.
Some elementary facts of the dimension theory of complex spaces are collected in section 3.1.

First Riemann removable singularity theorem. Assume X is a complex manifold and $A \subset X$ is a rare analytic subset.
The "weak form" of the theorem states:
If $f: X \to \mathbb{C}$ is continuous and holomorphic on $X \setminus A$, then f is holomorphic on X.
The "strong form" of the theorem states:
If $f: X \setminus A \to \mathbb{C}$ is holomorphic and $|f|$ is locally bounded on X, then there is a unique holomorphic extension $\hat{f}: X \to \mathbb{C}$ of f.

For proofs we refer to [57], [107], [150].

Second Riemann removable singularity theorem. Assume X is a complex manifold and $A \subset X$ is an analytic subset such that

$$\operatorname{codim}_x A \geq 2 \quad \text{for every } x \in X.$$

Then any holomorphic function $f: X \setminus A \to \mathbb{C}$ has a unique holomorphic extension $\hat{f}: X \to \mathbb{C}$.

We indicate a simple method of reducing the second theorem to the first theorem by using the maximum modulus principle. Obviously it is sufficient to prove the following

Lemma. Assume $X \subset \mathbb{C}^n$ is open and $A \subset X$ is analytic with $\dim A \leq n-2$. Then any holomorphic function

$$f: X \setminus A \to \mathbb{C}$$

is locally bounded on X.

For the proof of the lemma we use the

Remark. For $r > 0$ consider the ball $B := \{x \in \mathbb{C}^2: \|x\| < r\}$, an open neighbourhood U of $\overline{B}$ and a finite set $\{x_1,\ldots,x_n\} \subset B$. If

$$f: U \setminus \{x_1,\ldots,x_n\} \to \mathbb{C}$$

is holomorphic, then for any $p \in B \setminus \{x_1,\ldots,x_n\}$ we have

$$|f(p)| \leq \max\{|f(x)|: x \in \partial B\}.$$

Proof of the remark. Assume the assertion is false. Now choose a complex line L through p which does not meet $x_1,\ldots,x_n$. Then $L \cap \overline{B}$ is isomorphic to a one-dimensional closed disc and the maximum modulus principle for $f|(L \cap \overline{B})$ yields a contradiction.

Proof of the lemma. If $p \in X$ we have to find a neighbourhood U of p such that $f|(U \setminus A)$ is bounded.

If $(x_1,\ldots,x_n)$ are coordinates in $\mathbb{C}^n$ we put

$$X' := \{x \in X: x_1 = x_2 = 0\} \quad \text{and} \quad X'' := \{x \in X: x_3 = \ldots = x_n = 0\}.$$

Then we may assume $p = 0$ and that there is a ball B around 0 in X'' and a relatively compact open neighbourhood U of 0 in X'' such that

$$\overline{B} \times U \subset X,$$

$$A \cap (\overline{U} \times \partial B) = \emptyset \quad \text{and}$$

$$A \cap (\{x'\} \times \partial B) \quad \text{is finite for any } x' \in U.$$

This is an immediate consequence of the assumption $\dim A \leq n-2$ (see e.g. [ ]). Now it is sufficient to prove

$$|f(x)| \le \max\{|f(y)|: y \in \overline{U} \times \partial B\}$$

for any $x \in (B \times U) \setminus A$. But this follows from the above remark.

<u>2.24.</u> The Riemann removable singularity theorems do not hold on complex spaces. We give two standard examples:

<u>Example 1.</u> Consider the reduced complex subspace

$$X = \{(x_1,x_2) \in \mathbb{C}^2: x_1{}^2 - x_2{}^3 = 0\} \hookrightarrow \mathbb{C}^2$$

and the holomorphic map

$$\varphi: \mathbb{C} \to X,\ z \mapsto (z^3,z^2).$$

Obviously φ is a homeomorphism and it is easy to see that

$$f := \varphi^{-1}: X \to \mathbb{C}$$

is not holomorphic in o. But f is continuous and holomorphic outside o and equal to the restriction to X of x_1/x_2 .

<u>Example 2.</u> Consider the reduced complex subspace

$$X = \{(x_1,x_2) \in \mathbb{C}^2: (x_1 - x_2)x_2 = 0\} \hookrightarrow \mathbb{C}^2 \quad \text{and define}$$

$$f: X \setminus \{0\} \longrightarrow \mathbb{C}\ .\ \ x \longmapsto \frac{x_2}{x_1}\ .$$

f attains the values 1 and 0 on the two components of X; hence f cannot be extended holomorphically to the origin.

In order to characterize those complex spaces where the different forms of the Riemann removable singularity theorem hold we give the following definition:

Let X be a <u>reduced</u> complex space and denote by $S(X) \subset X$ the singular locus (see 2.15, 2.16).

f is called a *weakly holomorphic function on* X, if

a) f is defined on $X \setminus S(X)$ and $f: X \setminus S(X) \to \mathbb{C}$ is holomorphic

b) f is locally bounded on X.

f is called *continuous weakly holomorphic on* X, if

a) $f: X \to \mathbb{C}$ is continuous

b) f is holomorphic on $X \setminus S(X)$.

If we neglect formal troubles, a continuous weakly holomorphic function is weakly holomorphic.

Sometimes weakly holomorphic functions are defined to be holomorphic outside an arbitrary rare analytic subset. We show that this makes no difference.

<u>Remark.</u> Let X be a reduced complex space, $A \subset X$ an analytically rare set and

$$f: X \setminus A \longrightarrow \mathbb{C}$$

a holomorphic function which is locally bounded on X. Then there is a uniquely determined weakly holomorphic function f' on X which coincides with f outside $A \cup S(X)$.

<u>Proof.</u> Since $A \cap (X \setminus S(X))$ is analytically rare in the complex manifold $X \setminus S(X)$ we obtain f from the strong form of the first Riemann theorem (2.23).

For any reduced complex space X we define

$$\hat{\mathcal{O}}_X \subset \mathcal{C}_X$$

as the *sheaf of continuous weakly holomorphic functions* and

$$\tilde{\mathcal{O}}_X$$

as the *sheaf of weakly holomorphic functions.*

Clearly $\hat{\mathcal{O}}_X$ and $\tilde{\mathcal{O}}_X$ are $\mathcal{O}_X$-modules.

<u>Definition.</u> Let X be a reduced complex space, $p \in X$. X is called

$$\textit{maximal} \text{ in } p \iff \mathcal{O}_{X,p} = \hat{\mathcal{O}}_{X,p} ,$$

$$\textit{normal} \text{ in } p \iff \mathcal{O}_{X,p} = \tilde{\mathcal{O}}_{X,p} .$$

X is called maximal (resp. normal) if it is maximal (resp. normal) in every point.
The reader should be aware that maximality or normality in one point does not immediately imply the same property in a neighbourhood.

The complex space of Example 1 is not maximal in 0, in Example 2 it is not normal in 0.

<u>2.25.</u> The first important result on weakly holomorphic functions is the existence of "universal denominators".

<u>Proposition.</u> Let X be a reduced complex space. Then for any $p \in X$ there is an open neighbourhood U and a function $h \in \mathcal{O}_X(U)$ with the following

properties:

a) $h_x \in \mathcal{O}_{X,x}$ is a non zero divisor for any $x \in U$

b) $h_x \cdot \widetilde{\mathcal{O}}_{X,x} \subset \mathcal{O}_{X,x}$ for any $x \in U$.

Condition b) means that any $f_x \in \widetilde{\mathcal{O}}_{X,x}$ has a representation

$$f_x = \frac{g_x}{h_x} \quad \text{with } g_x \in \mathcal{O}_{X,x}$$

where h_x is "universal" for all f_x (i.e. a *universal denominator*).

For a proof of this proposition we refer to [107].

Corollary. Let X be a reduced complex space, $p \in X$. Then $\widetilde{\mathcal{O}}_{X,p}$ is a finite $\mathcal{O}_{X,p}$-module and equal to the integral closure of $\mathcal{O}_{X,p}$ in its full ring of quotients.

For the simple proof we refer to [107].

Recall that a ring without zero-divisors is called *normal*, if it is integrally closed in its full ring of quotients. Hence X is normal in p if and only if $\mathcal{O}_{X,p}$ is normal. This implies that $\mathcal{O}_{X,p}$ is an integral domain (see [50]). Consequently every normal complex space is locally irreducible. We leave it to the reader to derive this property directly from the definition.

Another important consequence of the existence of universal denominators is the

Lemma. For any reduced complex space X the *non-normal locus*

$$N(X) := \{x \in X: X \text{ is not normal in } x\} \subset X$$

is analytically rare.

A very simple and tricky proof of the fact that N(X) is analytic has been given by GRAUERT and REMMERT (see [107], p. 121). Since N(X) is contained in the singular locus S(X) it is rare (2.16).

2.26. For a reduced complex space X a holomorphic map

$$\nu: \widetilde{X} \to X$$

is called a *normalization* if the following conditions are satisfied:

a) $\widetilde{X}$ is a normal complex space.

b) ν is finite and surjective.

c) $A := \nu^{-1}N(X) \subset \widetilde{X}$ is analytically rare and the restriction

$$\nu' : \tilde{X} \smallsetminus A \longrightarrow X \smallsetminus N(X)$$

of ν is biholomorphic.
Here $N(X) \subset X$ denotes the non-normal locus (2.25).

Proposition. The normalization is uniquely determined up to an isomorphism, i.e. if

$$\nu_1 : \tilde{X}_1 \to X \quad \text{and} \quad \nu_2 : \tilde{X}_2 \to X$$

are two normalizations of X then there is a unique biholomorphic map $\Phi: \tilde{X}_1 \to \tilde{X}_2$ such that the diagram

$$\begin{array}{ccc} \tilde{X}_1 & \xrightarrow{\Phi} & \tilde{X}_2 \\ & {\scriptstyle\nu_1}\searrow \quad \swarrow{\scriptstyle\nu_2} & \\ & X & \end{array}$$

commutes.

Proof. Put

$$A_i := \nu_i^{-1} N(X) \quad \text{and} \quad \nu_i' := \nu_i | \tilde{X}_i \smallsetminus A_i \quad \text{for } i = 1,2.$$

Then by definition of a normalization

$$\Phi' := \nu_2'^{-1} \circ \nu_1' : \tilde{X}_1 \smallsetminus A_1 \longrightarrow \tilde{X}_2 \smallsetminus A_2$$

is a biholomorphic map. We have to extend Φ' to $\tilde{X}_1$. This problem being local with respect to $\tilde{X}_1$ we fix a point $x \in A_1$. If $p := \nu_1(x)$ we know

$$\nu_2^{-1}(p) = \{y_1, \ldots, y_k\}$$

and if $U \subset X$ is a sufficiently small open neighbourhood of p, then

$$\nu_2^{-1}(U) = W_1 \cup \ldots \cup W_k$$

where W_j is a neighbourhood of y_j for $j = 1, \ldots, k$ and this union is disjoint (see 1.10). Furthermore we may assume

$$W_1 \cup \ldots \cup W_k \hookrightarrow Z \subset \mathbb{C}^n$$

is a closed complex subspace where Z is a bounded open subset.
If $V \subset \tilde{X}_1$ is an open neighbourhood of x such that $\nu_1(V) \subset U$, then

$$\Phi' | U \smallsetminus A_1 : U \smallsetminus A_1 \longrightarrow W_1 \cup \ldots \cup W_k$$

is given by n bounded holomorphic functions on $U \smallsetminus A_1$. Since $\tilde{X}_1$ is normal they may be extended holomorphically to U. This yields a holomorphic extension

$$\Phi : \tilde{X}_1 \to \tilde{X}_2$$

of Φ'. Since A_2 is rare this extension is unique and for symmetry reasons Φ is biholomorphic.

Now we are able to prove the main result of this section (see [73]).

<u>Normalization-Theorem.</u> If X is a reduced complex space then the sheaf $\tilde{\mathcal{O}}_X$ of weakly holomorphic functions is a coherent $\mathcal{O}_X$-module and

$$\tilde{X} := \text{Specan } \tilde{\mathcal{O}}_X \xrightarrow{\ \nu\ } X$$

is a normalization of X. Moreover there is a canonical isomorphism

$$\tilde{\mathcal{O}}_X \longrightarrow \nu_* \mathcal{O}_{\tilde{X}} .$$

For every $p \in X$ there is a canonical correspondence of irreducible components of X at p and points of $\nu^{-1}(p)$.

In particular, if X is irreducible at p, then ν is topological at p, i.e. there is an open neighbourhood $U \subset X$ of p such that

$$\nu|\nu^{-1}(U): \nu^{-1}(U) \longrightarrow U$$

is topological.

We shall make use of the following well known algebraic facts.

<u>Lemma 1.</u> Let R be a reduced noetherian ring with minimal prime ideals $\mathfrak{P}_1, \ldots, \mathfrak{P}_m$. Then

$$\tilde{R} = \prod_{i=1}^{m} \widetilde{R/\mathfrak{P}_i} .$$

<u>Lemma 2.</u> If the local analytic $\mathbb{C}$-algebra R is an integral domain, then $\tilde{R}$ is local (see [107] IV, Prop.2, or [50], II §7, Satz 3).

<u>Lemma 3.</u> If $S_1, \ldots, S_m$ are local rings with maximal ideals $m_1, \ldots, m_m$, then

$$S := \prod_{i=1}^{m} S_i$$

is a semilocal ring. Its maximal ideals are given by

$$n_j := S_1 \times \ldots \times S_{j-1} \times m_j \times S_{j+1} \times \ldots \times S_m$$

and there is a canonical isomorphism of local rings

$$S_j \longrightarrow S_{n_j}$$

for $j = 1, \ldots, m$.

<u>Proof of the normalization theorem.</u> Because of the uniqueness of the normalization it is sufficient to give a local solution. So fix $p \in X$. From the corollary of 2.24 we know that $\tilde{\mathcal{O}}_{X,p}$ is a finite $\mathcal{O}_{X,p}$-module.

Hence there is an exact sequence

$$\mathcal{O}_{X,p}^{\ell} \xrightarrow{M_p} \mathcal{O}_{X,p}^{k} \longrightarrow \tilde{\mathcal{O}}_{X,p} \longrightarrow 0 ,$$

where M_p is determined by a $k\times\ell$-matrix of germs. If we shrink X we may replace the germs by holomorphic functions on X and the corresponding matrix M determines an exact sequence of $\mathcal{O}_X$-modules

$$\mathcal{O}_X^{\ell} \xrightarrow{M} \mathcal{O}_X^{k} \longrightarrow \mathcal{A} \longrightarrow 0 ,$$

where $\mathcal{A}$ is defined as the cokernel of M. By definition of $\mathcal{A}$ there is an isomorphism

$$\mathcal{A}_p \longrightarrow \tilde{\mathcal{O}}_{X,p} .$$

Hence the coherent $\mathcal{O}_X$-module $\mathcal{A}$ inherits the structure of an $\mathcal{O}_X$-algebra for sufficiently small X (use the fact that $\mathcal{A} \otimes_{\mathcal{O}_X} \mathcal{A}$ is coherent and 0.8). Now we define

$$\nu\colon \tilde{X} := \operatorname{Specan} \mathcal{A} \longrightarrow X$$

as the analytic spectrum with its canonical projection (for details see 1.12 - 1.13). In particular there is a canonical isomorphism

$$\mathcal{A} \longrightarrow \nu_* \mathcal{O}_{\tilde{X}} .$$

We first claim that $\tilde{X}$ is normal in every point $q \in \nu^{-1}(p)$. Again by assuming X sufficiently small we have a decomposition

$$X = X_1 \cup \ldots \cup X_m ,$$

where $X_1,\ldots,X_m$ are the irreducible components of X. If $\wp_i \subset \mathcal{O}_{X,p}$ is the ideal corresponding to X_i, then $\wp_i$ is a minimal prime ideal. By Lemma 1 we know

$$\mathcal{A}_p \cong \tilde{\mathcal{O}}_{X,p} = \prod_i \widetilde{\mathcal{O}_{X,p}/\wp_i} .$$

On the other hand the points of $\nu^{-1}(p)$ correspond to the maximal ideals of the semi-local ring $\mathcal{A}_p$. As the integral closure of local rings the rings $\widetilde{\mathcal{O}_{X,p}/\wp_i}$ are local (Lemma 2). Hence we may apply Lemma 3 and obtain

$$\nu^{-1}(p) = \{q_1,\ldots,q_m\} \quad \text{and} \quad \mathcal{O}_{\tilde{X},q_i} \cong \widetilde{\mathcal{O}_{X,p}/\wp_i} \quad \text{for } i = 1,\ldots,m$$

since the local rings of the analytic spectrum are the localizations of $\mathcal{A}_p$ (see 1.13). This immediately implies that $\tilde{X}$ is normal in the points $q_1,\ldots,q_m$. The set of normal points of $\tilde{X}$ being open (2.25) and ν being finite we may once more shrink X and assume $\tilde{X}$ is normal.

In case X is irreducible at p we have $m = 1$. Since ν is proper it is topological at p.

Now we may construct an isomorphism

$$A \longrightarrow \tilde{\mathcal{O}}_X .$$

Take $h \in \mathcal{O}_X(X)$ such that h_x is a non-zero-divisor and universal denominator for every $x \in X$ (2.25). Hence we may consider

$$\tilde{\mathcal{O}}_X \subset \mathcal{O}_X\left[\frac{1}{h}\right]$$

as a subring. By definition of A multiplication with h_p yields a monomorphism

$$A_p \longrightarrow \mathcal{O}_{X,p} .$$

If we compose it with multiplication by h_p^{-1} we obtain our given isomorphism

$$A_p \longrightarrow \tilde{\mathcal{O}}_{X,p} \subset \mathcal{O}_{X,p}\left[\frac{1}{h_p}\right] .$$

Hence for sufficiently small X multiplication first by h and then by h^{-1} yields a monomorphism

$$\alpha\colon A \longrightarrow \mathcal{O}_X\left[\frac{1}{h}\right]$$

of $\mathcal{O}_X$-algebras. If $x \in X$ and $\nu^{-1}(x) = \{y_1,\dots,y_n\}$ then

$$A_x = \prod_j \mathcal{O}_{\tilde{X},y_j}$$

is integrally closed since it is a product of integrally closed rings. Consequently $\alpha_x(A_x) = \tilde{\mathcal{O}}_{X,x}$ and α induces the desired isomorphism

$$A \longrightarrow \tilde{\mathcal{O}}_X$$

of $\mathcal{O}_X$-modules. In particular $\tilde{\mathcal{O}}_X$ is coherent.

Though it is not necessary for this proof let us extract a more explicit description of $\tilde{X}$ from the construction of the analytic spectrum. We may assume we have

$$f_1,\dots,f_k \in \tilde{\mathcal{O}}_X(X)$$

such that for every $x \in X$ their germs at x generate $\tilde{\mathcal{O}}_{X,x}$ over $\mathcal{O}_{X,x}$. Then there is an epimorphism of $\mathcal{O}_X$-algebras

$$\alpha\colon \mathcal{O}_X[t_1,\dots,t_k] \longrightarrow \tilde{\mathcal{O}}_X \quad \text{with } t_i \longmapsto f_i$$

and

$$\tilde{X} = \text{Specan } \tilde{\mathcal{O}}_X \hookrightarrow X \times \mathbb{C}^k$$

is a closed complex subspace. If $x \in X \setminus N(X)$ then $\tilde{\mathcal{O}}_{X,x} = \mathcal{O}_{X,x}$ and the kernel of

$$\alpha_x: \mathcal{O}_{X,x}[t_1,..,t_k] \longrightarrow \tilde{\mathcal{O}}_{X,x}$$

is generated by $t_1 - f_{1,x},..., t_k - f_{k,x}$. Consequently outside N(X) the normalization is the image of the embedding

$$\psi: X\setminus N(X) \longrightarrow (X\setminus N(X)) \times \mathbb{C}^k, \quad x \longmapsto (x, f_1(x),..,f_k(x)).$$

Since N(X) is rare in X the Lemma of RITT (0.44) implies that $\tilde{X}$ is the closure of the image of ψ in $X \times \mathbb{C}^k$.

We finally prove that

$$\nu: \text{Specan}\, \tilde{\mathcal{O}}_X \longrightarrow X$$

is a normalization. As we have already proved, X is normal, ν is finite and surjective. Furthermore we just have seen that outside N(X) the analytic spectrum is the image of an embedding of the base. Hence, if $A := \nu^{-1}(N(X))$, then the restriction of ν

$$\tilde{X}\setminus A \longrightarrow X\setminus N(X)$$

is biholomorphic. Of course this assertion may also be derived from the universal property of the analytic spectrum and the fact that $\tilde{\mathcal{O}}_{X,x} = \mathcal{O}_{X,x}$ for $x \in X\setminus N(X)$.

It remains to show that $A \subset \tilde{X}$ is rare. Take $y \in A$ and put $x := \nu(y)$. From the construction of $\tilde{X}$ we know that there is an open neighbourhood $U \subset X$ of x with the following properties: there is an irreducible component U' of U with corresponding prime ideal $\mathfrak{p} \subset \mathcal{O}_{X,x}$ such that

$$\mathcal{O}_{\tilde{X},y} = \widetilde{\mathcal{O}_{X,x}/\mathfrak{p}}$$

and the composition of the homomorphisms

$$\mathcal{O}_{X,x} \longrightarrow \mathcal{O}_{X,x}/\mathfrak{p} \hookrightarrow \widetilde{\mathcal{O}_{X,x}/\mathfrak{p}}$$

is just the canonical homomorphism

$$\tilde{\nu}_y: \mathcal{O}_{X,x} \longrightarrow \mathcal{O}_{\tilde{X},y}.$$

Hence there is a neighbourhood $V \subset \tilde{X}$ of y and a factorization

$$V \xrightarrow{\varphi} U' \hookrightarrow U$$

of $\nu|V$. Since U' is irreducible, φ is topological. We know $N(X) \subset X$ is rare, consequently N(X) is rare in U' and A is rare in V. This completes the proof of the normalization theorem.

<u>2.27.</u> As we have seen the normality of a complex space means that the strong form of the first Riemann removable singularity theorem is valid. As we shall see this implies that the second removable singularity theorem holds. We first prove a basic

<u>Integrity Lemma.</u> Assume X is a reduced pure-dimensional complex space, Y is a connected normal complex space and

$$\varphi: X \to Y$$

is a finite surjective holomorphic map. Then the canonical map $\mathcal{O}_Y(Y) \to \mathcal{O}_X(X)$ is injective (hence we may consider $\mathcal{O}_Y(Y) \subset \mathcal{O}_X(X)$ as a subring) and every $f \in \mathcal{O}_X(X)$ is integral over $\mathcal{O}_Y(Y)$, i.e. there are $a_1,\ldots,a_n \in \mathcal{O}_Y(Y)$ such that

$$f^n + a_1 f^{n-1} + \ldots + a_n = 0 .$$

In addition the number n may be choosen independent of f.

<u>Proof.</u> In every irreducible component of X there is a point p such that $\operatorname{cork}_p \varphi = 0$. Hence the set

$$M = \{x \in X: \operatorname{cork}_x \varphi > 0\} \subset X$$

is analytically rare (see 2.17) and this easily implies that

$$B := S(Y) \cup \varphi(S(X) \cup M) \subset Y \quad \text{and} \quad A := \varphi^{-1}(B) \subset X$$

are analytically rare. The definition of A and B together with 2.4. implies that

$$\varphi|(X \setminus A): X \setminus A \to Y \setminus B$$

is finite and locally biholomorphic. Consequently the number n of elements of $\varphi^{-1}(y)$ is the same for all $y \in Y \setminus B$, since $Y \setminus B$ stays connected.

Now we define holomorphic functions

$$\overline{\sigma_1},\ldots,\overline{\sigma_n}: Y \setminus B \to \mathbb{C} \quad \text{by}$$

$$\overline{\sigma_1}(y) := f(x_1) + \ldots + f(x_n),$$

$$\overline{\sigma_2}(y) := f(x_1)\cdot f(x_2) + \ldots + f(x_{n-1})\cdot f(x_n),$$

$$\vdots$$

$$\overline{\sigma_n}(y) := f(x_1)\cdot\ldots\cdot f(x_n),$$

if $\{x_1,\ldots,x_n\} = \varphi^{-1}(y)$. The identity

$$t^n - \overline{\sigma_1}(y)t^{n-1} + \ldots \pm \overline{\sigma_n}(y) = (t - f(x_1))\cdot\ldots\cdot(t - f(x_n))$$

for the elementary symmetric functions implies

$$\overline{f}^n - \overline{\sigma}_1 \overline{f}^{n-1} + \ldots \pm \overline{\sigma}_n = 0 \,, \quad \text{where} \quad \overline{f} := f|(X \setminus A) \,.$$

Since f is holomorphic on X the functions $\overline{\sigma}_1,\ldots,\overline{\sigma}_n$ have to be locally bounded on Y. Y being normal they may be extended to holomorphic functions on Y and since A is rare we have

$$f^n - \sigma_1 f^{n-1} + \ldots \pm \sigma_n = 0.$$

Corollary. Let X be a pure dimensional reduced complex space and assume the singular locus S(X) has codimension at least 2. Then every $f \in \mathcal{O}_X(X \setminus S(X))$ is weakly holomorphic on X.

Proof. We have to show that f is locally bounded on X. If $m := \dim X$ then for every $p \in X$ we may find an open neighbourhood $U \subset X$ of p, an open set $V \subset \mathbb{C}^m$ and a finite surjective holomorphic map

$$\varphi \colon U \to V$$

(see 3.1). The invariance of the dimension under finite holomorphic maps (see 3.2 and 3.8) implies

$$\dim \varphi(U \cap S(X)) \le m - 2 \,.$$

Put $B := \varphi(U \cap S(X))$ and $A := \varphi^{-1}(B)$. Then the restriction

$$\varphi' \colon U \setminus A \to V \setminus B$$

of φ is again finite. The Integrity Lemma implies that $g := f|U \setminus A$ is integral over $\mathcal{O}_Y(V \setminus B)$, i.e. there are $b_1,\ldots,b_n \in \mathcal{O}_Y(V \setminus B)$ such that

$$g^n + b_1 g^{n-1} + \ldots + b_n = 0 \,. \qquad (*)$$

By the second Riemann removable singularity theorem (2.23) there are holomorphic extensions $a_1,\ldots,a_n$ of $b_1,\ldots,b_n$. By shrinking U and V we may assume $|a_1|,\ldots,|a_n|$ are bounded on V. The relation (*) implies that $|g|$ is bounded on $U \setminus A$. Hence $|f|$ is bounded on $U \setminus S(X)$.

Lemma. In every normal complex space X the singular locus S(X) has codimension at least 2 in every point.

For a proof of this basic fact we refer to [107]. We derive the following interesting consequence, which finally shows that normal complex spaces can be characterized without using the technical notion of weakly holomorphic functions.

Proposition. For a complex space X the following conditions are equivalent:

i) X is normal

ii) For every open $U \subset X$ the restriction map

$$\mathcal{O}_X(U) \to \mathcal{O}_X(U \setminus S(X))$$

is bijective.

Proof. i) ⇒ ii). Since X is reduced the restriction maps are injective (2.16). X being normal the corollary and the lemma from above imply that they are surjective.
ii) ⇒ i). From 2.16 we know X is reduced. Then the normality is obvious.

These results together imply the following version of the

Second Riemann removable singularity theorem. Assume X is a normal complex space and $A \subset X$ is an analytic subset of codimension at least 2 in every point. Then every holomorphic function on $X \setminus A$ has a unique holomorphic extension to X.

In order to reformulate the above proposition we use the sets

$$S_k(\mathcal{O}_X) := \{p \in X: \text{prof } \mathcal{O}_{X,p} \le k\} \subset X$$

which are analytic for every complex space X and every k (see [122], [137]).

Theorem. For a complex space X the following conditions are equivalent:

i) X is normal
ii) $\dim(S(X) \cap S_k(\mathcal{O}_X)) \le k-2$ for all $k \ge 1$.

Proof. By a basic theorem on local cohomology (see e.g. [137], theorem 1.14) condition ii) is equivalent to condition ii) of the above proposition (see also [102]).

Corollary 1. For every complex space X the non-normal locus $N(X) \subset X$ is an analytic subset (see also 2.25).

Corollary 2. Assume X is Cohen-Macaulay (i.e. prof $\mathcal{O}_{X,p} = \dim_p X$ for every $p \in X$) and $\dim_p S(X) \le \dim_p X - 2$ for every p.
Then X is normal.

The simple proofs of the corollaries are left to the reader. Since in particular every hypersurface is Cohen-Macaulay we obtain the following

Example. The 2-dimensional cone

$$X = \{(x_1,x_2,x_3) \in \mathbb{C}^3: z_1{}^2 = z_2 z_3\} \subset \mathbb{C}^3$$

is a hypersurface and the origin is its singular locus. Hence it is normal.

2.28. For a holomorphic map $\varphi: X \to Y$ between reduced complex spaces a holomorphic map $\tilde{\varphi}: \tilde{X} \to \tilde{Y}$ such that the diagram

$$\begin{array}{ccc} \tilde{X} & \xrightarrow{\tilde{\varphi}} & \tilde{Y} \\ \downarrow & & \downarrow \\ X & \xrightarrow{\varphi} & Y \end{array}$$

commutes is called a *lifting to the normalizations*. In general such a lifting does not exist.

Example. Take a reduced complex space X such that the nonnormal locus N(X) is a normal complex space. Consider the normalization

$\nu: \tilde{X} \to X$

and assume the restriction of ν

$\nu^{-1} N(X) \to N(X)$

does not admit a holomorphic section. Then the canonical inclusion

$\iota: N(X) \to X$

cannot be lifted to the normalizations (Figure 5).

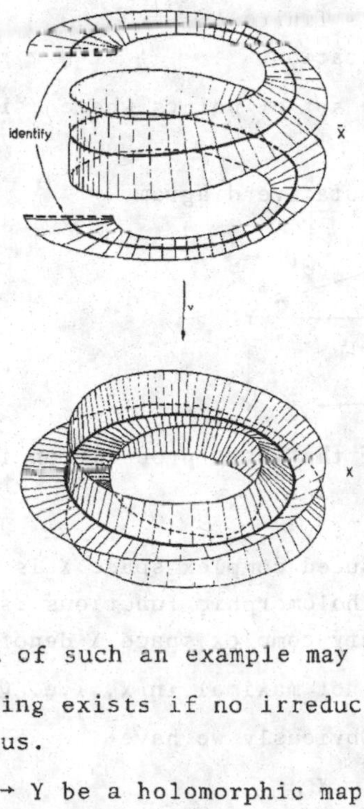

Figure 5

An explicit description of such an example may be found in 4.1.

Now we show that a lifting exists if no irreducible component is mapped into the non-normal locus.

Proposition. Let $\varphi: X \to Y$ be a holomorphic map between reduced complex spaces. If

$$\varphi^{-1}(N(Y)) \subset X$$

is rare, then there is a unique lifting of φ to the normalizations.

<u>Proof.</u> Denote by $\nu\colon \tilde{X} \to X$ the normalization map. The sets

$$A := \varphi^{-1}(N(Y)) \cup N(X) \subset X \quad \text{and} \quad \tilde{A} := \nu^{-1}(A) \subset \tilde{X}$$

are rare. Outside $\tilde{A}$ the lifting is given and since $\tilde{X}$ is normal standard arguments show that there is a unique holomorphic extension

$$\tilde{\varphi}\colon \tilde{X} \to \tilde{Y}\,.$$

Sometimes the following substitute for a lifting to the normalizations is useful.

<u>Lemma.</u> Let $\varphi\colon X \to Y$ be a holomorphic map between reduced complex spaces. Then there is a commutative diagram of holomorphic maps

$$\begin{array}{ccc} X' & \xrightarrow{\varphi'} & \tilde{Y} \\ {\scriptstyle\pi}\downarrow & & \downarrow \\ X & \xrightarrow{\varphi} & Y \end{array}$$

with the following properties:

a) X' is normal and π is finite

b) $\tilde{Y} \to Y$ is the normalization.

However, in general, the set of points where π is not biholomorphic is not a proper subset.

<u>Proof.</u> Look at the commutative diagram

$$\begin{array}{ccccc} X' & := & \widetilde{(X \times_Y \tilde{Y})}_{red} & \xrightarrow{\varphi'} & \\ & & \downarrow & & \\ & & (X \times_Y \tilde{Y})_{red} & \longrightarrow & \tilde{Y} \\ & {\scriptstyle\pi}\searrow & \downarrow & & \downarrow \\ & & X & \xrightarrow{\varphi} & Y\,. \end{array}$$

Even if the condition of the above proposition is satisfied X' need not be a normalization of X.

<u>2.29.</u> Recall that a reduced complex space X is maximal if the sheaf $\hat{\mathcal{O}}_X$ of continuous weakly holomorphic functions is equal to the structure sheaf $\mathcal{O}_X$. For an arbitrary complex space X denote by

$$M(X) = \{x \in X\colon X \text{ is not maximal in } x, \text{ i.e. } \mathcal{O}_{X,p} \neq \hat{\mathcal{O}}_{X,p}\}$$

the *non-maximal locus*. Obviously we have

$$M(X) \subset N(X) \subset S(X) \subset X\,,$$

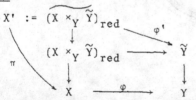

where $N(X)$ resp. $S(X)$ denote the non-normal respectively singular locus of X.

Now a holomorphic map $\mu\colon \hat{X} \to X$ is called a *maximalization* of X, if

a) $\hat{X}$ is maximal

b) μ is a homeomorphism

c) If $A := \mu^{-1}(M(X))$ then the restriction of μ

$$\hat{X} \setminus A \to X \setminus M(X)$$

is biholomorphic.

Maximalization Theorem. Let X be a reduced complex space. Then we have:

1) The sheaf $\hat{\mathcal{O}}_X$ is a coherent $\mathcal{O}_X$-module.
2) The non-maximal locus $M(X) \subset X$ is a rare analytic set.
3) The canonical holomorphic map

$$\mu\colon \hat{X} := \operatorname{Specan} \hat{\mathcal{O}}_X \;\to\; X$$

is a maximalization of X; in particular there is a canonical isomorphism

$$\hat{\mathcal{O}}_X \to \mu_* \mathcal{O}_{\hat{X}} .$$

4) There is a commutative diagram

$$\begin{array}{ccc} \tilde{X} & & \\ {\scriptstyle\nu}\downarrow & \searrow^{\psi} & \\ X & \xrightarrow{\ \mu\ } & \hat{X} \end{array}$$

of holomorphic maps, where ν denotes the normalization map.

5) If $\mu_1\colon \hat{X}_1 \to X$ and $\mu_2\colon \hat{X}_2 \to X$ are two maximalizations then there is a unique biholomorphic map $\Phi\colon \hat{X}_1 \to \hat{X}_2$ such that the diagram

$$\begin{array}{ccc} \hat{X}_1 & \xrightarrow{\ \Phi\ } & \hat{X}_2 \\ & {\scriptstyle\mu_1}\searrow \quad \swarrow{\scriptstyle\mu_2} & \\ & X & \end{array}$$

commutes.

Proof. We are going to derive the Maximalization Theorem from the Normalization Theorem via an equivalence relation. Implicitely this makes use of the finite coherence theorem. The key to the construction is the following obvious

Remark. If $\nu\colon \tilde{X} \to X$ denotes the normalization then for any $f \in \tilde{\mathcal{O}}_X(X) = \mathcal{O}_{\tilde{X}}(\tilde{X})$ the following conditions are equivalent:

i) $f \in \hat{\mathcal{O}}_X(X)$

ii) If $x \in X$ and $\nu^{-1}(x) = \{y_1, \dots, y_m\}$ then

$$f(y_1) = f(y_2) = \dots = f(y_m) .$$

Now we identify points of $\widetilde{X}$ with the same image in X. In order to be able to apply the machinery developed in chapter 1 we define

$$R := (\widetilde{X} \times_X \widetilde{X})_{red}$$

and since ν was proper we obtain from 1.22 a cokernel

$$R \rightrightarrows X \xrightarrow{\psi} X/R =: \hat{X} ,$$

where $\hat{X}$ is a complex space and ψ is holomorphic. The universal property of the cokernel implies the existence of a holomorphic map

$$\mu: \hat{X} \to X$$

such that $\nu = \mu \circ \psi$. By looking at the details of the construction in 1.22 we immediately see that $|\hat{X}| = |X|$ and $|\psi| = |\nu|$. Consequently

$$\hat{\mathcal{O}}_X = \mathcal{O}_{\hat{X}} \quad \text{and} \quad \hat{X} = \operatorname{Specan} \hat{\mathcal{O}}_X .$$

$\hat{\mathcal{O}}_X$ being a coherent $\mathcal{O}_X$-module we have

$$M(X) = \operatorname{supp}(\hat{\mathcal{O}}_X/\mathcal{O}_X).$$

This proves 1) and 2).

For 3) it remains to check that X is maximal. μ maps the singular locus of X into the singular locus of X. Hence a continuous weakly holomorphic function on an open subset $U \subset X$ is continuous weakly holomorphic on $\mu(U)$, consequently holomorphic on U.

Let us show 5). The map $\Phi := \mu_2^{-1} \circ \mu_1$ is topological and its restriction

$$\hat{X}_1 \setminus \mu_1^{-1}(M(X)) \to \hat{X}_2$$

is holomorphic. Since $\hat{X}_1$ is maximal Φ is holomorphic and for symmetry reasons even biholomorphic.

<u>2.30.</u> As we have seen in 2.28, in general a holomorphic map cannot be lifted to the normalizations. We are going to prove that such obstructions do not exist in the maximalizations.

We shall make use of the classical

<u>Theorem of CLEMENTS.</u> Assume $X \subset \mathbb{C}^n$ is open and $\varphi: X \to \mathbb{C}^n$ is holomorphic and injective. Then $\varphi(X) \subset \mathbb{C}^n$ is open and

$$\varphi: X \to \varphi(X)$$

is biholomorphic (see [26], [111] §20, [108] Chap.5).

<u>Lemma.</u> Let X,Y be reduced complex spaces and assume we have a holomorphic map $\varphi: X \to Y$ which is a homeomorphism. Then there is a rare analytic subset $A \subset Y$ such that the restriction of φ

$$X \setminus A \;\rightarrow\; Y \setminus \varphi(A)$$

is biholomorphic.

Proof. We define

$$A := S(X) \cup \varphi^{-1}S(Y) \ .$$

The singular loci being rare (2.16), A and $\varphi(A)$ are rare.

If $X' \longrightarrow X$ is an irreducible component then $\varphi(X') \longrightarrow Y$ is an irreducible complex subspace since φ is proper (see 1.18). φ being open, $\varphi(X')$ has to be an irreducible component of Y. As we shall see in Chapter 3 (3.2 and 3.8) we have

$$\dim X' = \dim \varphi(X').$$

Now the restriction of φ

$$\varphi' : X' \setminus A \longrightarrow \varphi(X') \setminus \varphi(A)$$

is a holomorphic homeomorphism between complex manifolds of equal dimension. The Theorem of CLEMENTS implies that φ' is biholomorphic and this proves the assertion.

Now we show that condition c) in the definition of a maximalization is superfluous.

Corollary. Let $\varphi: Z \rightarrow X$ be a holomorphic map between reduced complex spaces. If Z is maximal and φ is a homeomorphism then φ is a maximalization.

Proof. Let $\mu: \hat{X} \rightarrow X$ be a maximalization. The above lemma implies that there is a rare analytic set $A \subset X$ and a commutative diagram of biholomorphic maps

$$\begin{array}{ccc} \hat{X} \setminus \nu^{-1}(A) & \xrightarrow{\ \Phi'\ } & Z \setminus \varphi^{-1}(A) \\ & {\scriptstyle \nu'} \searrow \quad \swarrow {\scriptstyle \varphi'} & \\ & X \setminus A & \end{array}$$

where ν' and φ' are the restrictions of ν and φ. Since Φ' has a continuous extension $\Phi: \hat{X} \rightarrow Z$ the maximality of $\hat{X}$ and Z implies that Φ is biholomorphic (use the remark in 2.24).

Proposition. If $\varphi: X \rightarrow Y$ is a holomorphic map between reduced complex spaces then there is a unique holomorphic map $\hat{\varphi}: \hat{X} \rightarrow \hat{Y}$ such that the diagram

$$\begin{array}{ccc} \hat{X} & \xrightarrow{\hat{\varphi}} & \hat{Y} \\ \downarrow & & \downarrow \\ X & \xrightarrow{\varphi} & Y \end{array}$$

commutes. Here the vertical maps are the maximalizations.

Proof. Consider the commutative diagram

$$\begin{array}{ccc} \hat{Z} & \searrow^{\hat{\varphi}} & \\ \downarrow & & \\ (X \times_Y \hat{Y})_{red} =: Z & \longrightarrow & \hat{Y} \\ \downarrow & & \downarrow \\ X & \xrightarrow{\varphi} & Y . \end{array}$$

$\hat{Z}$ is maximal by definition and the composite map $\hat{Z} \to X$ is a holomorphic homeomorphism; hence the above corollary implies the assertion.

2.31. The results of 2.30 yield the following characterization of maximality.

Proposition. For a reduced complex space X the following conditions are equivalent:
i) X is maximal
ii) If Z is an arbitrary reduced complex space then every holomorphic homeomorphism $\varphi: Z \to X$ is biholomorphic.

Proof. i) $\Rightarrow$ ii). The composite holomorphic map

$$\hat{Z} \to Z \to X$$

is a homeomorphism and hence biholomorphic (see 2.30).
ii) $\Rightarrow$ i). The maximalization $\hat{X} \to X$ has to be biholomorphic.

As a simple consequence we obtain the

Graph Theorem. Assume $\varphi: X \to Y$ is a continuous map between reduced complex spaces and denote by

$$\Gamma_\varphi := \{(x,y) \in X \times Y : y = \varphi(x)\} \subset X \times Y$$

the graph of φ. If X is maximal the following conditions are equivalent:
i) φ is holomorphic
ii) $\Gamma_\varphi \subset X \times Y$ is an analytic subset.

Proof. i) $\Rightarrow$ ii). In case φ is holomorphic

$$\Gamma_\varphi = |X \times_Y Y| .$$

ii) $\Rightarrow$ i). Consider the holomorphic maps

$$\pi_1: \Gamma_\varphi \to X \quad \text{and} \quad \pi_2: \Gamma_\varphi \to Y$$

which are the restrictions of the canonical projections of $X \times Y$. φ being continuous π_1 is a homeomorphism and by the above proposition π_1 is biholomorphic. Hence

$$\varphi = \pi_2 \circ \pi_1^{-1}$$

is holomorphic.

Remark. The condition that φ be continuous cannot be dropped. This is easily seen from the following example: Put $X = Y = \mathbb{C}$ and

$$\varphi(z) = \begin{cases} \frac{1}{z} & \text{for } z \neq 0 \\ 0 & \text{for } z = 0 . \end{cases}$$

Then φ is not continuous but $\Gamma_\varphi \subset \mathbb{C} \times \mathbb{C}$ is an analytic subset; we have

$$\dim_o X \neq \dim_{(o,o)} \Gamma_\varphi .$$

It may easily be proved that in the above theorem the continuity condition may be replaced by the condition

$$\dim_p X = \dim_{(p,\varphi(p))} \Gamma_\varphi$$

for every $p \in X$ (see [117]).

2.32. We want to prove the following

Theorem. Let $\varphi: X \to Y$ be a finite holomorphic map. Then the following conditions are equivalent:

i) X is Stein
ii) Y is Stein.

The implication ii) $\Rightarrow$ i) has already been proved in 1.15. In order to show the converse we use a method of NARASIMHAN [106].

Lemma. Let $\psi: X \to Y$ be a proper surjective holomorphic map between reduced complex spaces and assume Y is normal. Then we have:

a) X holomorphically convex $\Rightarrow$ Y holomorphically convex.
b) X Stein $\Rightarrow$ Y Stein.

Proof. a) We only consider the case where Y has countable topology. Take a Stein factorization

$$X \xrightarrow{\rho} Z \xrightarrow{\zeta} Y$$

of φ (see 1.24). Since we have a canonical isomorphism

$$\mathcal{O}_Z(Z) \to \mathcal{O}_X(X) ,$$

Z is holomorphically convex (use 1.23).

If (p_i) is a sequence of points in Y without point of accumulation then there is a sequence of points (q_i) in Z without point of accumulation such that $p_i = \zeta(q_i)$ for all i. Choose $f \in \mathcal{O}_Z(Z)$ such that $(|f(q_i)|)$ is unbounded. We obviously may assume Z is pure dimensional and Y is connected. The Integrity Lemma of 2.27 implies that there are $a_1,..,a_n \in \mathcal{O}_Y(Y)$ such that

$$f^n + a_1 f^{n-1} + \ldots + a_n = 0 .$$

Hence there is at least one $\nu \in \{1,\ldots,n\}$ such that $(|a_\nu(p_i)|)$ is unbounded.

b) From a) we know Y is holomorphically convex. By the corollary of 1.25 it suffices to prove that every compact analytic subset $A \subset Y$ is finite. But this follows from the fact that $\varphi^{-1}(A) \subset X$ must be finite.

<u>Proof of the theorem.</u> We still have to show i) $\Rightarrow$ ii). By Corollary 1 of 0.37 we may assume X and Y are reduced. Put

$$n := \sup\{\dim_y Y\colon y \in Y\} .$$

In case $n < \infty$ we proceed by induction on n. For $n = 0$ the assertion is trivial. So assume it is proved for all complex spaces of dimension less than n.

Now consider the commutative diagram

$$\begin{array}{ccc} (X \times_Y \tilde{Y})_{red} & \longrightarrow & \tilde{Y} \\ \downarrow & & \downarrow \nu \\ X & \xrightarrow{\varphi} & Y \end{array}$$

of finite holomorphic maps where ν is the normalization. The first part of our theorem and the above lemma imply that $\tilde{Y}$ is Stein. In order to see that Y is Stein we prove

$$H^1(Y,\mathcal{J}) = 0$$

for every coherent ideal $\mathcal{J} \subset \mathcal{O}_Y$. We consider the coherent ideal

$$\mathcal{D} := \mathcal{O}_Y : \tilde{\mathcal{O}}_Y \subset \mathcal{O}_Y \quad \text{with } \operatorname{supp}(\mathcal{O}_Y/\mathcal{D}) = N(Y) \subset Y$$

(see 0.41), where N(Y) is the non-normal locus. Denote by

$$\mathcal{K} \subset \mathcal{O}_{\tilde{Y}}$$

the sheaf of ideals which is the image of

$$\nu^*(\mathcal{D}\cdot\mathcal{J}) = \varphi^{-1}(\mathcal{D}\cdot\mathcal{J}) \otimes_{\varphi^{-1}\mathcal{O}_Y} \mathcal{O}_{\tilde{Y}} \longrightarrow \mathcal{O}_{\tilde{Y}}$$

under the multiplication. We know

$$\nu_* K \subset \nu_* \mathcal{O}_{\tilde{Y}} = \tilde{\mathcal{O}}_Y \supset \mathcal{O}_Y \supset J$$

and by using the definition of K and the remark in 1.14 it is easy to check that

$$\mathcal{J} := \nu_* K \subset J .$$

K is just the ideal of the inverse image of the subspace of Y defined by $\mathcal{D}\cdot J$ (see 0.27). $\mathcal{J}$ is coherent since ν was finite and $\tilde{Y}$ being Stein we have (see 1.16)

$$H^1(Y,\mathcal{J}) = H^1(\tilde{Y},K) = 0. \qquad (*)$$

Since ν is biholomorphic outside N(Y) we know

$$\mathrm{supp}(J/\mathcal{J}) \subset N(Y).$$

$\nu^{-1}N(Y) \subset \tilde{Y}$ is Stein and the restriction of ν

$$\nu^{-1}N(Y) \to N(Y)$$

is finite. The dimension of N(Y) is less than n, hence the induction hypothesis implies that N(Y) is Stein and consequently

$$H^1(X,J/\mathcal{J}) = H^1(N(Y),J/\mathcal{J}) = 0. \qquad (**)$$

If we pass from the exact sequence

$$0 \to \mathcal{J} \to J \to J/\mathcal{J} \to 0$$

to the exact cohomology sequence, (*) and (**) imply

$$H^1(X,J) = 0.$$

If Y is not finite dimensional we use an exhaustion. Define $Y_k \hookrightarrow Y$ as the union of all irreducible components of dimension not greater than k. Then

$$Y_o \subset Y_1 \subset \ldots \subset \bigcup_{k=o}^{\infty} Y_k = Y.$$

Since X has countable topology (0.37), Y has countable topology and we may check holomorphic convexity with sequences. If

$$A := \{q_o, q_1, \ldots\}$$

with pairwise different q_i has no point of accumulation we construct $f \in \mathcal{O}_Y(Y)$ with $f(q_i) = i$.
Obviously there is an $f_o \in \mathcal{O}_{Y_o}(Y_o)$ with $f_o(q_i) = i$ for all i with $q_i \in Y_o$. Assume we already have constructed a holomorphic $f_k: Y_k \to \mathbb{C}$ with $f_k(q_i) = i$ for all i with $q_i \in Y_k$. As we already know Y_{k+1} is Stein, the closed subspace

$$Y_k \cup (A \cap Y_{k+1}) \hookrightarrow Y_{k+1}$$

is Stein. By Theorem B (0.37) we may find a holomorphic $f_{k+1}: Y_{k+1} \to \mathbb{C}$

such that $f_{k+1}|Y_k = f_k$ and $f_{k+1}(q_i) = i$ for all i with $q_i \in Y_{k+1}$.

Two different points of Y are contained in some Y_k. They can be separated by some f_k holomorphic in Y_k and by similar arguments as above they can be separated by some f holomorphic in Y. This completes the proof of the theorem.

Corollary. For a reduced complex space X the following conditions are equivalent:

i) X is Stein.

ii) The maximalization $\hat{X}$ is Stein.

iii) The normalization $\tilde{X}$ is Stein.

★ ★

CHAPTER 3. DEGENERACIES AND FLATNESS

In chapter 3 we had stated regularity conditions for holomorphic maps which can be tested with differential criterions. In this chapter we are going to consider continuity conditions for the fibres of a holomorphic map.

3.1. We shall frequently need basic facts from dimension theory of complex spaces. Let us therefore recall some well known results. If X is a complex space and $p \in X$ we define the dimension of X in p as

$$\dim_p X := \dim \mathcal{O}_{X,p} ,$$

i.e. the *Krull-dimension* of its local ring at p.

Theorem 1. Let X be a complex space. Then for $p \in X$ and $k \in \mathbb{N}$ the following conditions are equivalent:

i) $\dim_p X \leq k$

ii) There is a finite homomorphism of local $\mathbb{C}$-algebras

$$\rho: \mathbb{C}[\langle z_1,\ldots,z_k\rangle] \longrightarrow \mathcal{O}_{X,p} .$$

iii) There are polydiscs $Z' \subset \mathbb{C}^k$ and $Z'' \subset \mathbb{C}^m$, an open neighbourhood U of p in X and an embedding

$$\iota: U \longrightarrow Z' \times Z''$$

with the following property: If

$$\pi_1 : Z' \times Z'' \longrightarrow Z'$$

denotes the canonical projection on the first factor then

$$\pi_1 \circ \iota: U \longrightarrow Z'$$

is finite (see 1.10).

In addition, $\dim_p X = k$ is equivalent to the fact that in ii) we may choose ρ finite and injectivo or in iii) we may choose $\pi_1 \circ \iota$ finite and surjective (see e.g. [118], [107], [50], [150]).

Corollary 1 (Semicontinuity of the dimension). If X is a complex space then every $p \in X$ has a neighbourhood U such that

$$\dim_x X \leq \dim_p X \quad \text{for all } x \in U.$$

This follows immediately from condition iii).

We state another important result which may be proved in the same context.

<u>Theorem 2</u> (Open mapping and maximum modulus principle). Let X be an irreducible complex space and $f: X \to \mathbb{C}$ a non-constant holomorphic function. Then

a) f is an open map

b) $|f|$ has no local maximum.

Thereby f is called *constant* if it has constant values, i.e. $f(x) = f(y)$ for all $x,y \in X$.

<u>Corollary 2.</u> If X is a compact connected complex space then every holomorphic function $f: X \to \mathbb{C}$ is constant.

<u>Proof.</u> $|f|$ attains a maximum on every irreducible component of X. Then f is constant on every irreducible component and, X being connected, everywhere.

It should be noted that

$$n := \dim_{\mathbb{C}} \mathcal{O}_X(X) < \infty$$

if X is compact by the Finiteness Theorem 1.17. But in general $n > 1$ unless X is reduced and connected.

<u>Corollary 3.</u> Every compact and holomorphically separable complex space X is finite.

<u>Proof.</u> Since X is compact it has a finite number of connected components. Every component is compact, connected and holomorphically separable. By Corollary 2 it consists of one point.

<u>3.2.</u> Let us call a holomorphic map $\varphi: X \to Y$ *finite in* $p \in X$, if p is an isolated point of its fibre $X_{\varphi(p)}$. By using a simple topological argument we prove that finite <u>in</u> p implies finite <u>at</u> p.

<u>Lemma.</u> Let $\varphi: X \to Y$ be a holomorphic map which is finite in p. Then there are open neighbourhoods $U \subset X$ of p and $V \subset Y$ of $\varphi(p)$ such that

$$\varphi|U: U \to V$$

is finite.

<u>Proof.</u> Lemma 1 of 1.10 yields a proper $\varphi|U: U \to V$. We may assume U and consequently the fibres of $\varphi|U$ are closed complex subspaces of an open subset in some $\mathbb{C}^n$. Hence Corollary 3 of 3.1 implies that $\varphi|U$ has finite fibres.

Now we can prove an important consequence. Recall that a continuous map

$\varphi: X \to Y$ is called *open in* $p \in X$ if for every neighbourhood U of p in X the image $\varphi(U)$ is a neighbourhood of $\varphi(p)$ in Y. Note that this does not imply that φ is open in a neighbourhood of p.

Theorem. Let $\varphi: X \to Y$ be a holomorphic map which is finite in $p \in X$. Then there are arbitrary small open neighbourhoods U of p in X such that:

a) $\varphi(U)$ is analytic at $\varphi(p)$.

b) $\dim_{\varphi(p)}\varphi(U) = \dim_p X$.

c) $\varphi|U: U \to \varphi(U)$ is open in p.

Proof. The above lemma yields open neighbourhoods U of p and V of $q := \varphi(p)$ such that

$$\varphi|U: U \to V$$

is finite. Hence by REMMERTs mapping theorem $\varphi(U) \subset V$ may be provided with a canonical structure of a complex space. If U is such that $U \cap \varphi^{-1}(q) = \{p\}$ then according to the definition of $\mathcal{O}_{\varphi(U)}$ the canonical homomorphism

$$\alpha: \mathcal{O}_{\varphi(U),q} \to (\varphi_*\mathcal{O}_U)_q = \mathcal{O}_{X,p}$$

is injective. Since $\varphi_*\mathcal{O}_U$ is a coherent $\mathcal{O}_V$-module, α is finite. If $k := \dim_q \varphi(U)$ then there is a finite injective homomorphism

$$\beta: \mathbb{C}[\langle z_1,\ldots,z_k\rangle] \to \mathcal{O}_{\varphi(U),q} .$$

Now $\alpha\circ\beta$ is finite and injective; this proves b). Since we may choose U arbitrarily small, a) and b) imply c).

Let us finally show that $\varphi|U$ need not be open:

Example. Consider the closed complex subspace $X \hookrightarrow \mathbb{C}^3$ which is the union of the plane

$$X_1 = \{(z_1,z_2,z_3): z_3 = 0\}$$

and the line

$$X_2 := \{(z_1,z_2,z_3): z_2 = 0,\ z_1 - z_3 = 0\} .$$

The holomorphic map

$$\varphi: X \to \mathbb{C}^2$$

which is the restriction of the projection

$$\mathbb{C}^3 \to \mathbb{C}^2,\ (z_1,z_2,z_3) \mapsto (z_1,z_2),$$

is finite in 0 but not open in any neighbourhood of 0.

Corollary. If the holomorphic map $\varphi: X \to Y$ is finite and open in $p \in X$, then

$$\dim_p X = \dim_{\varphi(p)} Y .$$

3.3. In 0.35 we have seen that locally a holomorphic map may be embedded. Now we prove the existence of a local *spreading*.

Proposition. If $\varphi: X \to Y$ is a holomorphic map and $p \in X$, put $k := \dim_p X_{\varphi(p)}$. Then there is an open neighbourhood U of p in X and a commutative diagram of holomorphic maps

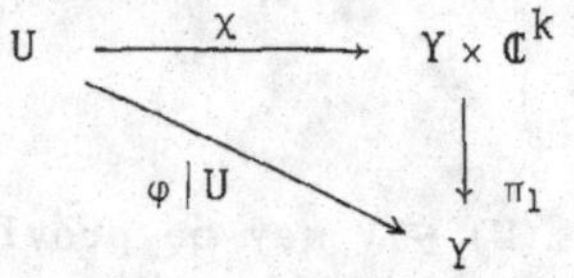

(*)

such that χ is finite in p.

Proof. By assumption there is an open neighbourhood U' of p in the fibre $X_{\varphi(p)}$ and a holomorphic map

$$\psi: U' \to \mathbb{C}^k$$

which is finite in p. For a sufficiently small neighbourhood U of p in X there is a holomorphic map

$$\Psi: U \to \mathbb{C}^k$$

generating ψ. If we define

$$\chi := (\varphi|U, \Psi): U \to Y \times \mathbb{C}^k$$

then obviously the diagram (*) commutes. Since

$$X_{\chi(p)} = (U \cap X_{\varphi(p)}) \cap X_{\Psi(p)}$$

χ is finite in p and the assertion is proved.

3.4. As an easy consequence we obtain a *semicontinuity* statement for the fibre dimension.

Proposition. Let $\varphi: X \to Y$ be a holomorphic map. Then every point $p \in X$ has a neighbourhood U in X such that

$$\dim_x X_{\varphi(x)} \le \dim_p X_{\varphi(p)} \quad \text{for all } x \in U.$$

Proof. Put $k := \dim_p X_{\varphi(p)}$. By 3.2 and 3.3 we may find open neighbourhoods U of p in X, V of $\varphi(p)$ in Y, an open subset $Z \subset \mathbb{C}^k$ and a commutatative diagram of holomorphic maps

$$\begin{array}{ccc} U & \xrightarrow{\chi} & V \times Z \\ & \searrow^{\varphi|U} & \downarrow \pi_1 \\ & & V \end{array}$$

such that χ is finite. Hence for any $x \in U$ the restriction of χ

$$X_{\varphi(x)} \cap U \quad \rightarrow \quad \{x\} \times Z$$

is finite and the assertion follows from 3.1.

3.5. In order to prove that the fibre dimension is even Zariski-semi-continuous we shall use differential devices. Recall that in 2.17 we had defined a differential corank and rank for a holomorphic map $\varphi: X \rightarrow Y$ at $p \in X$ by

$$\mathrm{cork}_p \varphi = \dim_{\mathbb{C}} T_p X_{\varphi(p)} = \mathrm{emdim}_p X_{\varphi(p)}$$

$$\mathrm{rk}_p \varphi = \dim_{\mathbb{C}} T_p X - \dim_{\mathbb{C}} T_p X_{\varphi(p)} = \mathrm{emdim}_p X - \mathrm{emdim}_p X_{\varphi(p)} \quad .$$

Analogously we introduce a *geometric corank* and *rank* at p by (the reader may forgive us the clumsy notation)

$$\mathrm{dimcork}_p \varphi := \dim_p X_{\varphi(p)} \quad \text{and}$$

$$\mathrm{dimrk}_p \varphi := \dim_p X - \dim_p X_{\varphi(p)} \quad .$$

As in the differential case the corank will turn out to be more practicable (unless the complex spaces X and Y are pure dimensional). This concept has been studied systematically by REMMERT [117]. Furthermore we introduce the global coranks

$$\mathrm{cork}\ \varphi := \min\{\mathrm{cork}_p \varphi : p \in X\} \quad \text{and}$$

$$\mathrm{dimcork}\ \varphi := \min\{\mathrm{dimcork}_p \varphi : p \in X\}$$

which are nothing but the *minimal embedding dimension* (resp. *minimal dimension*) of the fibres.

Since every embedding is a finite holomorphic map we know

$$\dim_p X \leq \mathrm{emdim}_p X$$

for every complex space X and every $p \in X$. Consequently

$$\mathrm{dimcork}_p \varphi \leq \mathrm{cork}_p \varphi \quad \text{and}$$

$$\mathrm{dimcork}\ \varphi \leq \mathrm{cork}\ \varphi$$

for every holomorphic map $\varphi: X \rightarrow Y$ and every $p \in X$.

Analogously to the differential singular locus (see 2.17)

$$\mathrm{Sing}^k\varphi = \{p \in X : \mathrm{cork}_p\varphi > k\}$$

we define for every $k \in \mathbb{N}$ the *degeneracy set of order* k by

$$\mathrm{Deg}^k\varphi := \{p \in X : \mathrm{dimcork}_p\varphi > k\} .$$

We obviously have

$$\mathrm{Deg}^k\varphi \subset \mathrm{Sing}^k\varphi$$

for every k. With this bunch of definitions in mind we may state the following simple

Proposition. Let $\varphi: X \to Y$ be a holomorphic map and assume X is reduced. Then

$$\mathrm{dimcork}\ \varphi = \mathrm{cork}\ \varphi$$

and for every $k \in \mathbb{N}$ we have

$$\mathrm{Deg}^k\varphi = X \iff \mathrm{Sing}^k\varphi = X .$$

Proof. Put

$$d := \mathrm{dimcork}\ \varphi \quad \text{and} \quad c := \mathrm{cork}\ \varphi .$$

We already know $d \leq c$. To prove the opposite inequality let us assume $d < c$. Then take a point $p \in X$ such that

$$d = \mathrm{dimcork}_p\varphi .$$

By the semicontinuity of the fibre dimension (3.4) there is an open neighbourhood U of p in X such that

$$d = \mathrm{dimcork}_x\varphi \quad \text{for all } x \in U . \tag{*}$$

Now we have

$$c' := \min\{\mathrm{cork}_x\varphi : x \in U\} \geq c \quad \text{and}$$

$$A := \{x \in U : \mathrm{cork}_x\varphi > c'\} \subset U$$

is a proper analytic subset. If we take $q \in U \setminus A$, then by Corollary 2 of 2.19 our map φ is a c'-mersion at q, hence

$$\mathrm{dimcork}_q\varphi = c' \geq c > d .$$

This contradicts (*) and we conclude $d = c$.
Finally the obvious equivalences

$$\mathrm{Deg}^k\varphi = X \iff k < d \quad \text{and}$$

$$\mathrm{Sing}^k\varphi = X \iff k < c$$

imply the stated equivalence.

3.6. Now we can prove a result due to CARTAN [18] and REMMERT [117] on the *analyticity of the degeneracy sets*. Here we adopt the simplified method of HOLMANN [70] which makes use of elementary differential concepts on complex spaces and does not need the REMMERT-STEIN-THULLEN-Theorem on essential singularities of analytic sets.

Theorem. If $\varphi: X \to Y$ is a holomorphic map then for every $k \in \mathbb{N}$ the degeneracy set

$$\mathrm{Deg}^k\varphi = \{x \in X : \dim_x X_{\varphi(x)} > k\} \subset X$$

is analytic.

Proof. Since the dimension of every complex space is equal to the dimension of its reduction we may assume X is reduced. Now consider the diagram of reduced complex spaces

$$\begin{array}{ccccccccc} X & \supset & \mathrm{Sing}^k\varphi_o & \supset & \mathrm{Sing}^k\varphi_1 & \supset \ldots \supset & \mathrm{Sing}^k\varphi_n & \supset \ldots \\ & & \cup & & \cup & & \cup & \\ \mathrm{Deg}^k\varphi & = & \mathrm{Deg}^k\varphi_o & = & \mathrm{Deg}^k\varphi_1 & = \ldots = & \mathrm{Deg}^k\varphi_n & = \ldots , \end{array}$$

where we recursively define $\varphi_o := \varphi$ and

$$\varphi_{n+1} := \varphi|\mathrm{Sing}^k\varphi_n : \mathrm{Sing}^k\varphi_n \to Y .$$

By definition we have the inclusions in the first line. Furthermore we know there are squares

$$\begin{array}{ccc} \mathrm{Sing}^k\varphi_n & \supset & \mathrm{Sing}^k\varphi_{n+1} \\ \cup & & \cup \\ \mathrm{Deg}^k\varphi_n & \supset & \mathrm{Deg}^k\varphi_{n+1} . \end{array}$$

Now take $x \in \mathrm{Deg}^k\varphi_n$ and denote by $X' \subset \mathrm{Sing}^k\varphi_{n-1}$ the fibre through x of

$$\varphi_n : \mathrm{Sing}^k\varphi_{n-1} \to Y .$$

Then $\dim_x X' > k$. Hence there is an irreducible component $X'' \subset X'$ of dimension greater than k containing x and obviously $X'' \subset \mathrm{Sing}^k\varphi_n$. If we now consider

$$\varphi_{n+1} : \mathrm{Sing}^k\varphi_n \to Y$$

we see that $x \in \mathrm{Deg}^k\varphi_{n+1}$. This finally proves

$$\mathrm{Deg}^k\varphi_n = \mathrm{Deg}^k\varphi_{n+1} .$$

If $U \subset X$ is a relatively compact open subset then

$$\mathrm{Sing}^k\varphi_n \cap U = \mathrm{Sing}^k\varphi_{n+1} \cap U$$

for some n. Proposition 3.5 implies

$$\operatorname{Sing}^k \varphi_{n+1} \cap U = \operatorname{Deg}^k \varphi_{n+1} \cap U = \operatorname{Deg}^k \varphi \cap U$$

and the assertion is proved.

3.7. We show that the image of a holomorphic map with constant fibre dimension is a locally analytic set. This is an easy consequence of the following

Factorization Lemma. Let $\varphi: X \to Y$ be a holomorphic map between reduced complex spaces and assume

$$\operatorname{dimcork}_x \varphi := \dim_x X_{\varphi(x)} = k$$

for all $x \in X$. Then for every $p \in X$ there is a commutative diagram

$$\begin{array}{ccc} U & \xrightarrow{\ \chi\ } & V' \times Z \\ {\scriptstyle \varphi|U}\downarrow & & \downarrow{\scriptstyle \pi_1} \\ V & \xleftarrow{\ \iota\ } & V' \end{array}$$

of holomorphic maps, where $U \subset X$ and $V \subset Y$ are open neighbourhoods of p and $\varphi(p)$, $Z \subset \mathbb{C}^k$ is a domain, ι is a closed embedding and χ is finite and surjective.

Proof. By 3.2 and 3.3 we obtain the required neighbourhoods U and V, a domain $Z \subset \mathbb{C}^k$ and a finite holomorphic map

$$\chi: U \to V \times Z$$

which is a spreading of $\varphi|U$. By REMMERTs Mapping Theorem $\chi(U) \hookrightarrow V \times Z$ is a closed complex subspace. If $\chi(p) = (\varphi(p), 0)$ we define

$$V' := \chi(U) \cap (V \times \{0\})$$

and we obtain a canonical embedding $\iota: V' \hookrightarrow V$. Now it suffices to prove

$$\chi(U) = V' \times Z .$$

If $y \in \varphi(U)$ then the restriction of χ

$$\chi_y: X_y \cap U \to \{y\} \times Z$$

is again finite. Since Z is connected χ_y is surjective for every $y \in \varphi(U)$ by 3.2 and this implies

$$\chi(U) = V' \times Z .$$

Corollary. Let $\varphi: X \to Y$ be a holomorphic map between reduced complex spaces and assume

$$\dim_x X_{\varphi(x)} = k$$

for all $x \in X$. Then for every $p \in X$ there are arbitrarily small open neighbourhoods $U \subset X$ of p and $V \subset Y$ of $\varphi(p)$ with the following properties:

a) $\varphi(U) \subset V$ is an analytic subset.

b) The "dimension formula"

$$\dim_{\varphi(p)}\varphi(U) = \dim_p X - \dim_p X_{\varphi(p)} =: \operatorname{dimrk}_p\varphi$$

holds.

c) $\varphi|U: U \to \varphi(U)$ is open in p.

Proof. We choose U and V as in the Factorization Lemma. Then a) is a consequence of

$$\varphi(U) = \iota(V').$$

Since χ is finite and open in p we obtain c) and furthermore (see 3.2)

$$\dim_p X = \dim_{\chi(p)}(V' \times Z) = \dim_{\varphi(p)}\varphi(U) + k ;$$

this proves b).

In 1.18 we had provided the image of a proper holomorphic map with a canonical complex structure. If the fibre dimension is constant this does not work in general.

Example. Consider the complex plane $\mathbb{C}$ with the coordinate function z. For every natural number $n \geq 1$ denote by

$$\iota_n: X_n := (\{0\}, \mathbb{C}[\langle z\rangle]/(z^n)) \hookrightarrow \mathbb{C}$$

the canonical embedding of the n-fold origin. If

$$X := \bigcup_{n\geq 1} X_n$$

is the disjoint union we define

$$\iota: X \to \mathbb{C} \quad \text{by} \quad \iota|X_n := \iota_n .$$

Then the kernel of

$$\tilde{\iota}: \mathcal{O}_{\mathbb{C}} \to \iota_*\mathcal{O}_X$$

is not of finite type.

3.8. In 3.4 we had defined the geometric rank of a holomorphic map $\varphi: X \to Y$ at $x \in X$ by

$$\operatorname{dimrk}_x\varphi = \dim_x X - \dim_x X_{\varphi(x)} .$$

Now a global geometric rank of φ is given by

$$\operatorname{dimrk}\varphi := \sup\{\operatorname{dimrk}_x\varphi : x \in X\} .$$

We want to show that this is a measure for the dimension of the image of φ. It should be noted that there are alternative definitions of a geometric rank (see e.g. [4]).

Lemma. Let $\varphi: X \to Y$ be a holomorphic map. Assume X is reduced and

$$X = \bigcup_{i\in I} X_i$$

is the decomposition into irreducible components. If $\varphi_i := \varphi|X_i$ then

$$\operatorname{dimrk} \varphi = \sup\{ \operatorname{dimrk} \varphi_i : i \in I \}.$$

Proof. For any $x \in X$ take a component $X_j \hookrightarrow X$ with $\dim_x X_j = \dim_x X$. Then

$$\dim_x X - \dim_x X_{\varphi(x)} \le \dim_x X_j - \dim_x (X_{\varphi(x)} \cap X_j)$$

and consequently

$$\operatorname{dimrk}_x \varphi \le \max\{ \operatorname{dimrk}_x \varphi_i : i \in I \text{ with } x \in X_i \} .$$

This immediately implies

$$\operatorname{dimrk} \varphi \le \sup\{ \operatorname{dimrk} \varphi_i : i \in I \} .$$

In order to prove the opposite inequality consider an arbitrary component $X_i \hookrightarrow X$. By using theorem 3.6 we find a point $p \in X_i$ such that X is irreducible in p and

$$\operatorname{dimrk} \varphi_i = \operatorname{dimrk}_p \varphi_i = \operatorname{dimrk}_p \varphi .$$

This proves the assertion.

Theorem. Let $\varphi: X \to Y$ be a proper holomorphic map, where X is reduced and irreducible. Put

$$r := \operatorname{dimrk} \varphi \quad \text{and} \quad D := \{x \in X : \operatorname{dimrk}_x \varphi < r \} .$$

Then

$$\dim \varphi(X) = r \quad \text{and} \quad \dim \varphi(D) \le r - 2 .$$

Proof. Recall that $D = \{x \in X: \operatorname{dimcork}_x \varphi > \operatorname{dimcork} \varphi\}$ since X is irreducible, hence D is analytic by 3.6. In order ot prove the theorem we proceed by induction on $m := \dim X$.

For $m = 0$ we have $r = 0$ and $D = \emptyset$. Since φ is proper, $\varphi(X)$ is discrete and consequently $\dim \varphi(X) = 0$.

Since X is irreducible and $D \subset X$ is a proper subset we know

$$\dim D \le m - 1 .$$

Take an irreducible component $D' \subset D$ and put $\varphi' := \varphi|D'$, $r' := \operatorname{dimrk} \varphi'$. We claim

$$r' \le r - 2 . \qquad (*)$$

There is a point $x \in D'$ such that D is irreducible in x and

$$\operatorname{dimrk}_x \varphi' = r' .$$

Since $x \in D$ we know $\dim_x X_{\varphi(x)} \geq m - r + 1$. Hence there is an irreducible component $X' \subset X_{\varphi(x)}$ with $x \in X'$ and

$$\dim X' = \dim_x X_{\varphi(x)} .$$

This implies $X' \subset D$, hence $X' \subset D'$, hence $X' \subset D'_{\varphi'(x)}$. Now the inequalities

$$\dim D' \leq m - 1 \quad \text{and} \quad \dim D'_{\varphi'(x)} \geq m - r + 1$$

imply (*). The induction hypothesis implies

$$\dim \varphi'(D') \leq r - 2 .$$

Since the decomposition of D into irreducible components is locally finite and φ was proper this implies

$$\dim \varphi(D) \leq r - 2 .$$

If $p \in X \setminus D$ the dimension formula of Corollary 3.7 implies

$$\dim_{\varphi(p)} \varphi(X) \geq r .$$

Hence $\varphi(D) \subset \varphi(X)$ is a proper analytic subset and it is sufficient to prove

$$\dim_q \varphi(X) = r \qquad (**)$$

for $q \in \varphi(X) \setminus \varphi(D)$. If we again use 3.7 and the assumption that φ is proper we may find an open neighbourhood $V \subset Y$ of q such that

$$\varphi^{-1}(V) = U_1 \cup \ldots \cup U_k \quad \text{and} \quad \varphi(U_1) \subset V, \ldots, \varphi(U_k) \subset V$$

are analytic subsets of dimension r in q. This implies (**).

<u>Corollary.</u> Let $\varphi: X \to Y$ be a proper holomorphic map. Then

$$\dim \varphi(X) = \operatorname{dimrk} \varphi .$$

In particular $\dim \varphi(X) \leq \dim X$.

<u>Proof.</u> We may assume X is reduced and consider the decomposition

$$X = \bigcup_{i \in I} X_i$$

into irreducible components. If $\varphi_i := \varphi | X_i$ then $\dim \varphi(X_i) = \operatorname{dimrk} \varphi_i$ by the above theorem. φ being proper the union

$$\varphi(X) = \bigcup_{i \in I} \varphi(X_i)$$

is locally finite. Hence we have

$$\dim \varphi(X) = \sup\{\dim \varphi(X_i) : i \in I\}$$

and the assertion follows from the above lemma.

For the case of not necessarily proper holomorphic maps we state a result of REMMERT-STEIN [119].

Theorem. Let $\varphi: X \to Y$ be a holomorphic map between complex spaces and assume X has countable topology. Put

$$r := \operatorname{dimrk} \varphi .$$

Then $\varphi(X) \subset Y$ is metrisable subspace with countable topology and has topological dimension $2r$.

M. KNESER [89] has given an example of a bijective real-analytic map $\varphi: X \to Y$ where X and Y are real-analytic manifolds with uncountable topology, where $\dim X = 2$ and $\dim Y = 3$. It seems to be unknown if such examples are possible in complex-analytic geometry (see [17]).

3.9. A not completely trivial example for a non-open holomorphic map is a σ-modification (see 4.1). We want to show that the openness of a holomorphic map $\varphi: X \to Y$ is essentially equivalent to a "dimension formula"

$$\dim_p X = \dim_{\varphi(p)} Y + \dim_p X_{\varphi(p)} , \text{ i.e.}$$

$$\operatorname{dimrk}_p \varphi = \dim_{\varphi(p)} Y$$

for all $p \in X$ (see [117]).

Proposition. Let $\varphi: X \to Y$ be a holomorphic map between complex spaces and assume $p \in X$. Then

$$\operatorname{dimrk}_p \varphi \le \dim_{\varphi(p)} Y . \qquad (*)$$

If Y is irreduzible in $\varphi(p)$ and if

$$\operatorname{dimrk}_p \varphi = \dim_{\varphi(p)} Y \qquad (**)$$

then φ is open in p (see 3.2).

Proof. Put $k := \dim_p X_{\varphi(p)}$ and $r := \dim_{\varphi(p)} Y$. By 3.2 and 3.3 we may find open neighbourhoods $U \subset X$ of p and $V \subset Y$ of $\varphi(p)$ with holomorphic maps

$$\chi: U \longrightarrow V \times \mathbb{C}^k \quad \text{and} \quad \psi: V \longrightarrow \mathbb{C}^r$$

which are finite in p resp. $\varphi(p)$. Consequently the map

$$(\psi \times id_{\mathbb{C}^k})\ \chi : U \longrightarrow \mathbb{C}^{r+k}$$

is finite in p and this implies

$$\dim_p X \le r + k \ .$$

This proves (*). Let us assume U and V are connected and choose a domain $Z \subset \mathbb{C}^k$ such that

$$\chi : U \longrightarrow V \times Z$$

is finite (use the lemma in 3.2). Since $V \times Z$ is irreducible (**) implies that χ is surjective and by the theorem in 3.2 χ is open in p. Since χ was a spreading φ is open in p.

Of course φ need not be open in a neighbourhood of p. As an example take a complex space X which is not pure dimensional in $p \in X$ and a holomorphic map $\varphi : X \to \mathbb{C}^k$ with $k = \dim_p X$ which is finite in p.

<u>Corollary 1.</u> Let $\varphi : X \to Y$ be a holomorphic map where Y is locally irreducible and assume

$$\operatorname{dimrk}_x \varphi = \dim_{\varphi(x)} Y$$

for every $x \in X$. Then φ is open.

<u>Proof.</u> The proposition implies that φ is open in every point. Then φ is open.

<u>Corollary 2.</u> Let $\varphi : X \to Y$ be a holomorphic map. Assume that X is pure-dimensional in a point $p \in X$, Y is irreducible in $\varphi(p)$ and

$$\operatorname{dimrk}_p \varphi = \dim_{\varphi(p)} Y \ .$$

Then φ is open in a neighbourhood of p.

<u>Proof.</u> The proposition implies

$$\dim_p X \le \dim_{\varphi(x)} Y + \dim_x X_{\varphi(x)}$$

for all $x \in X$. The semicontinuity of the fibre dimension together with our assumptions implies the opposite inequality and hence equality for all x in a neighbourhood of p. Now the assertion follows from Corollary 1.

<u>3.10.</u> Now we show that conversely the "dimension formula" holds for open holomorphic maps.

Recall that an element of a local analytic algebra is called *active* if it is a non-zerodivisor in the reduction (see [50]).

Lemma 1. Given a complex space X, a point $p \in X$, a holomorphic map $f: X \to \mathbb{C}$ with $f(p) = 0$, denote by $f_p \in m_{X,p}$ the germ of f at p and consider the following conditions:

i) f_p is active in $\mathcal{O}_{X,p}$

ii) f is open in p

iii) f is open in a neighbourhood of p.

Then we have the implications iii) $\Rightarrow$ i) $\Rightarrow$ ii). In case X is puredimensional in p we have i) $\Rightarrow$ iii).

Proof. If f_p is not active then the values of f are identically zero on a component of X_{red} passing through p. Then f is not open in any neighbourhood of p.
Now assume f_p is active. Then the so called "Active Lemma" (see [50]) implies

$$\dim_p X_o = \dim_p X - 1 .$$

By the corollaries of 3.9 we see that f is open in p and open in a neighbourhood of p in case X is puredimensional in p.
Obviously ii) is weaker than i).

Lemma 2. Let $\varphi: X \to Y$ be an open holomorphic map. Fix a point $p \in X$ and assume Y is pure-dimensional in $\varphi(p)$. Then for every active element $b \in m_{Y,\varphi(p)}$ the element

$$a := \tilde{\varphi}_p(b) \in m_{X,p}$$

is active.

Proof. If we replace X and Y by suitable open neighbourhoods of p and $\varphi(p)$ we obtain a commutative diagram of holomorphic maps

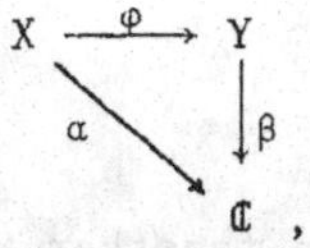

where β represents b and α represents a. By Lemma 1 we may assume β is open; hence α is open and a is active.

Proposition. Let $\varphi: X \to Y$ be an open holomorphic map and assume Y is pure-dimensional. Then

$$\operatorname{dimrk}_x \varphi = \dim_{\varphi(x)} Y$$

for every $x \in X$.

Proof. We proceed by induction on $n := \dim Y$. For $n = 0$ the assertion is trivial.

If $n \geq 1$ there is an active element $b \in m_{Y,\varphi(x)}$. By Lemma 2 the element $a := \tilde{\varphi}_x(b)$ is active. We may replace X and Y by suitable open neighbourhoods of x and $\varphi(x)$ and obtain a cartesian square

$$\begin{array}{ccc} X' & \hookrightarrow & X \\ \varphi' \downarrow & & \downarrow \varphi \\ Y' & \hookrightarrow & Y , \end{array}$$

where X' resp. Y' are the closed complex subspaces defined by representatives of a resp. b. It is easy to see that Y' is again pure dimensional and φ' is again open. Since a and b are active we know (see [50])

$$\dim_x X' = \dim_x X - 1 \quad \text{and} \quad \dim_{\varphi(x)} Y' = \dim_{\varphi(x)} Y - 1 .$$

Hence the induction hypothesis implies

$$\dim_x X' = \dim_{\varphi(x)} Y' + \dim_x X'_{\varphi(x)} .$$

Since $X'_{\varphi(x)} = X_{\varphi(x)}$ the assertion follows by adding 1 to both sides of this equality.

Simple examples show that the condition on the pure dimensionality of Y may not be dropped.

If we combine the above result with 3.9 we obtain:

<u>Theorem.</u> Let $\varphi: X \to Y$ be a holomorphic map between complex spaces and assume Y is locally irreducible. Then the following conditions are equivalent:

i) φ is open.

ii) $\dim_x X = \dim_{\varphi(x)} Y + \dim_x X_{\varphi(x)}$ for every $x \in X$.

3.11. In the second part of this chapter we want to study the concept of "flatness" of a holomorphic map. It is very difficult to interpret this algebraic notion in a geometric way. In [104] D. MUMFORD writes: "The concept of flatness is a riddle that comes out of algebra, but which technically is the answer to many prayers."

For an excellent introduction to flatness in analytic geometry we refer to [29].

Let us first recall a well known algebraic notion (see e.g. [5]).

Definition. Let R be a ring (as always commutative and with unit element). An R-module M is called *flat* (or more precisely R-flat) if it satisfies one of the following equivalent conditions:

i) For every exact sequence of R-modules

$$\dots \longrightarrow N_{i-1} \longrightarrow N_i \longrightarrow N_{i+1} \longrightarrow \dots$$

the induced sequence

$$\dots \longrightarrow N_{i-1} \otimes_R M \longrightarrow N_i \otimes_R M \longrightarrow N_{i+1} \otimes_R M \longrightarrow \dots$$

is again exact.

ii) For every short exact sequence of R-modules

$$0 \longrightarrow N' \longrightarrow N \longrightarrow N'' \longrightarrow 0$$

the induced sequence

$$0 \longrightarrow N' \otimes_R M \longrightarrow N \otimes_R M \longrightarrow N'' \otimes_R M \longrightarrow 0$$

is again exact.

iii) For every injective homomorphism of R-modules

$$N' \longrightarrow N$$

the induced homomorphism

$$N' \otimes_R M \longrightarrow N \otimes_R M$$

is again injective.

In analytic geometry this notion is used in the following situation:

Let $\varphi: X \to Y$ be a holomorphic map between arbitrary complex spaces and let $\mathcal{F}$ be an $\mathcal{O}_X$-module.

$\mathcal{F}$ is called *φ-flat in* $p \in X$, if $\mathcal{F}_p$ is $\mathcal{O}_{Y,\varphi(p)}$-flat.

φ is called *flat in* $p \in X$, if $\mathcal{O}_{X,p}$ is $\mathcal{O}_{Y,\varphi(p)}$-flat.

Thereby $\mathcal{O}_{X,p}$-modules are considered as $\mathcal{O}_{Y,\varphi(p)}$-modules via the canonical homomorphism

$$\tilde{\varphi}_p : \mathcal{O}_{Y,\varphi(p)} \longrightarrow \mathcal{O}_{X,p} .$$

$\mathcal{F}$ resp. φ are called *φ-flat* resp. *flat* if they have this property in every point of X.

A non-flat holomorphic map can be found in 0.10, Example 2 .

It is one aspect of flatness that it imposes strong continuity conditions upon the fibration induced by the given map. In the following sections we want to make this more precise. First we need some more algebraic machinery.

Remark. If $\Phi: R \to S$ is a ring homomorphism and M is a flat R-module, then $M \otimes_R S$ is a flat S-module.

Proof. For any S-module N we have isomorphisms

$$(M \otimes_R S) \otimes_S N \cong M \otimes_R (S \otimes_S N) \cong M \otimes_R N .$$

Hence any monomorphism of S-modules

$$N' \longrightarrow N$$

tensored over S with $M \otimes_R S$ yields a homomorphism of S-modules

$$M \otimes_R N' \longrightarrow M \otimes_R N .$$

Since M is R-flat it is an injective homomorphism of R-modules.

In algebraic geometry this remark suffices to prove that flatness is preserved by change of base. In analytic geometry the completed tensor product presents additional difficulties (see 3.15).

Let us recall the definition of "Tor" which may be used as a measure for non-flatness. We start with R-modules M and N and take any projective resolution (for instance a free resolution) of M

$$\ldots \quad F_{k+1} \to F_k \to F_{k-1} \to \ldots \to F_1 \to F_o \to M \to 0 .$$

If we tensor with N and leave off the last term at the right we obtain a complex

$$\ldots \longrightarrow F_{k+1} \otimes_R N \xrightarrow{d_{k+1}} F_k \otimes_R N \xrightarrow{d_k} F_{k-1} \otimes_R N \longrightarrow \ldots \xrightarrow{d_1} F_o \otimes_R N \xrightarrow{d_o} 0.$$

Now we define for $k \geq 0$

$$\mathrm{Tor}_k^R(M,N) := \mathrm{Ker}\, d_k \,/\, \mathrm{Im}\, d_{k+1} .$$

The right exactness of the tensor product yields an isomorphism

$$\mathrm{Tor}_o^R(M,N) \longrightarrow M \otimes N .$$

This definition of Tor has the following properties:

a) $\mathrm{Tor}_k^R(M,N)$ is independent of the choice of the projective resolution of M

b) There are isomorphisms

$$\mathrm{Tor}_k^R(M,N) \longrightarrow \mathrm{Tor}_k^R(N,M) \quad \text{for } k \geq 0$$

c) For any short exact sequence of R-modules

$$0 \longrightarrow N' \longrightarrow N \longrightarrow N'' \longrightarrow 0$$

there is a long exact Tor-sequence

$$\longrightarrow \mathrm{Tor}_k^R(N',M) \rightarrow \mathrm{Tor}_k^R(N,M) \rightarrow \mathrm{Tor}_k^R(N'',M) \rightarrow$$

$$\rightarrow \mathrm{Tor}_{k-1}^R(N',M) \rightarrow \mathrm{Tor}_{k-1}^R(N,M) \rightarrow \mathrm{Tor}_{k-1}^R(N'',M) \rightarrow$$

$$\cdots\cdots \qquad\qquad\qquad \cdots\cdots$$

$$\rightarrow \mathrm{Tor}_1^R(N',M) \rightarrow \mathrm{Tor}_1^R(N,M) \rightarrow \mathrm{Tor}_1^R(N'',M) \rightarrow$$

$$\rightarrow N' \otimes_R M \rightarrow N \otimes_R M \rightarrow N'' \otimes_R M \rightarrow 0 .$$

For a proof and more details we refer to [22]. In order to check flatness one can use the following

<u>Corollary.</u> For an R-module M the following conditions are equivalent:

i) M is R-flat

ii) $\mathrm{Tor}_k^R(M,N) = 0$ for every R-module N and $k \geq 1$

iii) $\mathrm{Tor}_1^R(M,N) = 0$ for every R-module N .

<u>3.12.</u> For later use we note two simple algebraic facts.

<u>Lemma.</u> Let $\Phi: R \rightarrow S$ be a ring homomorphism such that S is R-flat and fix $a \in R$.

1) $\Phi(a)$ zerodivisor $\Rightarrow$ a zerodivisor
2) Assume in addition that R, S and Φ are local. Then Φ is injective and consequently

   a zerodivisor $\Rightarrow$ $\Phi(a)$ zerodivisor .

Proof. If $\mathfrak{a} = \mathrm{Ann}(a)$ we have the exact sequence

$$0 \longrightarrow \mathfrak{a} \longrightarrow R \xrightarrow{a} R$$

where a denotes multiplication by a . Tensoring by S yields the exact sequence

$$0 \longrightarrow \Phi(\mathfrak{a})\cdot S \longrightarrow S \xrightarrow{\Phi(a)} S$$

where $\Phi(a)$ denotes multiplication by $\Phi(a)$. If $\Phi(a)$ is a zerodivisor we obtain $\Phi(\mathfrak{a}) \neq 0$, hence $\mathfrak{a} \neq 0$ and a is a zerodivisor. This proves 1).

If Φ is local we have $\Phi(m) \subset n$, where $m \subset R$ and $n \subset S$ denote the maximal ideals. If $a \neq 0$ then $1 \notin \mathfrak{a}$, hence $\mathfrak{a} \subset m$ and $\Phi(\mathfrak{a}) \subset n$. But if $\Phi(a) = 0$ then $\Phi(\mathfrak{a})\cdot S = S$. This implies that Φ is injective.

It follows easily from 3.11 that every free module is flat.

For finitely generated modules we prove:

Proposition. Let R be a local noetherian ring. If M is a finitely generated and flat R-module, then M is free.

Proof. If $m \subset R$ denotes the maximal ideal we put $k := R/m$. If $n := \dim_k M/mM$ NAKAYAMAs Lemma yields an exact sequence

$$0 \longrightarrow K \longrightarrow R^n \longrightarrow M \longrightarrow 0 .$$

By tensoring with k we obtain the exact sequence of k-vector spaces

$$0 = \mathrm{Tor}_1^R(M,k) \longrightarrow K \otimes_R k \longrightarrow R^n \otimes_R k \longrightarrow M \otimes_R k \longrightarrow 0 .$$

In particular we obtain

$$\dim_k K/mK = 0 \ , \text{ hence } K = 0 \ ,$$

again by NAKAYAMAs Lemma. This proves that M is free.

3.13. Proposition. For a finite holomorphic map $\varphi: X \to Y$ the following conditions are equivalent:

i) φ is flat
ii) $\varphi_*\mathcal{O}_X$ is locally free.

<u>Proof.</u> By 1.10 for every $q \in Y$ we have

$$(\varphi_* 0_X)_q = \prod_{x \in X_q} 0_{X,x} \, .$$

This product is free if and only if all factors $0_{X,x}$ are free. By 3.12 this is equivalent to the flatness of φ in all points of X_q .

This proposition admits a flatness test by "counting multiplicities". If $\varphi \colon X \to Y$ is finite then for any $q \in Y$ and $x \in X_q$ the local ring $0_{X_q,x}$ is a finite dimensional $\mathbb{C}$-vector space.
We define

$$\nu_x(\varphi) := \dim_{\mathbb{C}} 0_{X_q,x} \quad \text{and}$$

$$\nu_q(\varphi) := \sum_{x \in X_q} \nu_x(\varphi) \, .$$

<u>Corollary.</u> For a finite holomorphic map $\varphi \colon X \to Y$ with Y reduced, the following conditions are equivalent:

i) φ is flat

ii) $q \longmapsto \nu_q(\varphi)$ is a locally constant function on Y .

<u>Proof.</u> If $F := \varphi_* 0_X$ then by 1.8 F is locally free if and only if the function

$$q \longmapsto \dim_{\mathbb{C}} F_q / m_{Y,q} F_q$$

is locally constant on Y . Hence the isomorphisms

$$(\varphi_* 0_X)_q \otimes_{0_{Y,q}} \mathbb{C} \longrightarrow \prod_{x \in X_q} 0_{X,x} \otimes_{0_{Y,q}} \mathbb{C} \longrightarrow \prod_{x \in X_q} 0_{X_q,x}$$

imply the assertion.

We leave it as an exercise (look at the construction of the analytic spectrum in 1.15) to prove the following

<u>Remark.</u> If $\varphi \colon X \to Y$ is a finite holomorphic map then for every $q \in Y$ there is an open neighbourhood $V \subset Y$ and a commutative diagram

$$\begin{array}{ccc} U & \xrightarrow{\;\iota\;} & U' \\ {\scriptstyle \varphi|U}\downarrow & \swarrow {\scriptstyle \varphi'} & \\ V & & \end{array}$$

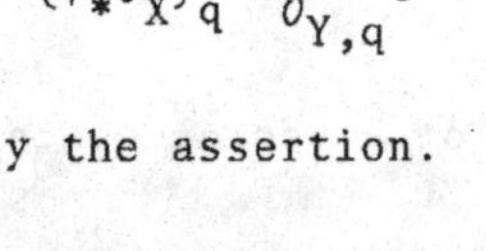

where $U = \varphi^{-1}(V)$, ι is an embedding, φ' is finite and flat.

Example (due to A. DOUADY). Put

$$Y = \{(p,q) \in \mathbb{C}^2 : 4p^3 + 27q^2 = 0\} \quad \text{and}$$

$$X = \{(p,q,x) \in \mathbb{C}^3 : 4p^3 + 27q^2 = 0\ ,\ x^3 + px + q = 0\}$$

and provide them with the complex structure defined by the given equations. Then Y is reduced but X has a simple and a double branch. The projection

$$\mathbb{C}^3 \longrightarrow \mathbb{C}^2\ ,\quad (p,q,x) \longmapsto (p,q)$$

restricts to a finite holomorphic map

$$\varphi : X \longrightarrow Y$$

and it is easy to check that

$$\nu_y(\varphi) = 3 \quad \text{for every } y \in Y.$$

Hence φ is flat. But now consider the restriction

$$\varphi_{red} : X_{red} \longrightarrow Y\ .$$

With a bit of computation one finds (X_{red} has two simple branches)

$$\nu_y(\varphi_{red}) = \begin{cases} 2 & \text{for } y \in Y \setminus \{0\}\ , \\ 3 & \text{for } y = 0\ . \end{cases}$$

Hence φ_{red} is not flat.

For another example with $Y = \mathbb{C}^2$ we refer to R.C. COWSIK and M.V. NORI: *On Cohen-Macaulay Rings*. Journal of Algebra 38 (1976), 536-538.

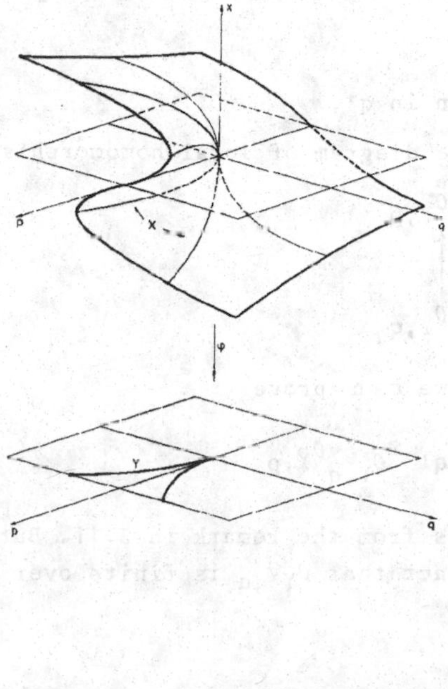

Figure 6

3.14. In order to derive flatness criterions for not necessarily finite holomorphic maps we need a fundamental algebraic result due to BOURBAKI-GROTHENDIECK (see [14] III, §5).

Theorem. Let $\Phi: R \to S$ be a local homomorphism between local noetherian rings and denote by $\mathfrak{m} \subset R$ the maximal ideal. If M is a finitely generated S-module and $\mathfrak{a} \subset \mathfrak{m} \subset R$ is an ideal, then the following conditions are equivalent:

i) M is R-flat

ii) $M/\mathfrak{a}M$ is $R/\mathfrak{a}$-flat and $\mathrm{Tor}_1^R(R/\mathfrak{a}, M) = 0$

iii) $M/\mathfrak{a}^k M$ is $R/\mathfrak{a}^k$-flat for $k \geq 1$

iv) $\mathrm{Tor}_1^R(R/\mathfrak{m}, M) = 0$.

3.15. Now we prove the basic property of flatness that it is preserved by change of base.

Proposition. Assume we have a cartesian square

$$\begin{array}{ccc} X' & \xrightarrow{\chi} & X \\ {\scriptstyle\varphi'}\downarrow & & \downarrow{\scriptstyle\varphi} \\ Y' & \xrightarrow{\psi} & Y \end{array}$$

of complex spaces and holomorphic maps. Take $p' \in X'$ and $p := \chi(p') \in X$. If φ is flat in p, then φ' is flat in p'.

Proof. We may apply the factorization of 3.3 to ψ in the point $q' := \varphi'(p')$. Hence it is sufficient to prove the proposition for the following special cases:

1) ψ is finite in q'

2) ψ is a submersion in q'.

Case 1. Consider the diagram of local homomorphisms

$$\begin{array}{ccc} \mathcal{O}_{X',p'} & \longleftarrow & \mathcal{O}_{X,p} \\ \uparrow & & \uparrow \\ \mathcal{O}_{Y',q'} & \longleftarrow & \mathcal{O}_{Y,q} \end{array}$$

where $q := \varphi(p)$. If we can prove

$$\mathcal{O}_{X',p'} \cong \mathcal{O}_{Y',q'} \otimes_{\mathcal{O}_{Y,q}} \mathcal{O}_{X,p} \qquad (*)$$

the assertion follows from the remark in 3.11. But (*) is a well known consequence of the fact that $\mathcal{O}_{Y',q}$ is finite over $\mathcal{O}_{Y,q}$ (see e.g. [9],

§1). A very short proof of (*) can be given by using the analytic spectrum. The question being local we may assume $\mathcal{B} := \psi_* \mathcal{O}_{Y'}$ is a coherent $\mathcal{O}_Y$-module and $\mathcal{B}_q = \mathcal{O}_{Y',q'}$. Then by assertion c) of 1.15 we have isomorphisms

$$\mathcal{O}_{X',p'} \longrightarrow (\varphi^*\mathcal{B})_p \longrightarrow \mathcal{O}_{Y',q'} \otimes_{\mathcal{O}_{Y,q}} \mathcal{O}_{X,p} .$$

<u>Case 2.</u> Translated into local algebra we have to prove the following assertion:
If $\Phi: R \to S$ is a flat homomorphism between local analytic algebras, then for any n the induced homomorphism

$$R[\langle t_1,..,t_n\rangle] \longrightarrow S[\langle t_1,..,t_n\rangle]$$

is flat.
Define $\mathfrak{a} := (t_1,..,t_n) \subset R[\langle t_1,..,t_n\rangle]$ and consider the following commutative diagram

$$\begin{array}{ccccc} S'/\mathfrak{a}^k S' & \longleftarrow & S' := S[\langle t_1,..,t_n\rangle] & \longleftarrow & S \\ \uparrow & & \uparrow & & \uparrow \\ R'/\mathfrak{a}^k & \longleftarrow & R' := R[\langle t_1,..,t_n\rangle] & \longleftarrow & R \end{array},$$

where all homomorphisms are the obvious ones. Since $R \longrightarrow R'/\mathfrak{a}^k$ is finite for $k \geq 1$, case 1 implies that $S'/\mathfrak{a}^k S'$ is $R'/\mathfrak{a}^k$-flat and now 3.14 implies that S' is R'-flat.

<u>Remark.</u> With the notations as above, assume that $\mathcal{F}$ is a coherent $\mathcal{O}_X$-module which is φ-flat in p. Then $\chi^*\mathcal{F}$ is φ'-flat in p'.
For a proof of this more general assertion we refer to [52], Exp. 13.

<u>3.16.</u> As another consequence of 3.14 we derive a flatness criterion for holomorphic maps $X \longrightarrow \mathbb{C}^n$ (see [80]).

Let M be a module over the local ring $(R,\mathfrak{m})$. A sequence $(a_1,..,a_n)$ of elements $a_i \in \mathfrak{m}$ is called an M-*sequence* if for $1 \leq i \leq n$ the multiplication homomorphism

$$a_i: M/(a_1M + \ldots + a_{i-1}M) \longrightarrow M/(a_1M + \ldots + a_{i-1}M)$$

is injective.
For instance if $R = \mathbb{C}[\langle t_1,..,t_n\rangle]$, then $(t_1,..,t_n)$ is an R-sequence.

Lemma. Let S be a local analytic $\mathbb{C}$-algebra and M a finitely generated S-module. If

$$\Phi: R = \mathbb{C}[\, t_1,..,t_n \,] \longrightarrow S$$

is a local homomorphism then the following conditions are equivalent:

i) M is R-flat

ii) $(t_1,..,t_n)$ is an M-sequence.

Proof. i) ⇒ ii). Put $R_i := R/(t_1R+...+t_{i-1}R)$ and $M_i := M/(t_1M+...+t_{i-1}M$ for $1 \le i \le n$. Since

$$M_i = M \otimes_R R_i$$

the remark of 3.11 implies that M_i is R_i-flat. Since the multiplication

$$t_i: R_i \to R_i$$

is injective, the multiplication

$$t_i: M_i \to M_i$$

is injective.

ii) ⇒ i). We proceed by induction on n. For $n = 0$ we have $R = \mathbb{C}$ and every M is R-flat because it is free.

For $n \ge 1$ put

$$R' = R/t_nR \quad \text{and} \quad M' := M/t_nM .$$

If $t'_1,..,t'_{n-1} \in R'$ denote the residue classes of $t_1,..,t_{n-1}$, then they are an M'-sequence. By induction hypothesis M' is R'-flat.

Now the free resolution

$$0 \longrightarrow R \xrightarrow{t_n} R \longrightarrow R' \longrightarrow 0$$

of R' tensored with M yields the exact sequence

$$0 \longrightarrow M \xrightarrow{t_n} M \longrightarrow M' \longrightarrow 0 ,$$

since t_n belongs to an M-sequence. This implies

$$\mathrm{Tor}_1^R(M,R') = 0$$

and by 3.15 we conclude that M is R-flat.

Corollary. Let $\varphi: X \to \mathbb{C}^n$ be a holomorphic map and take $p \in X$ with $\varphi(p) = 0$. Then the following conditions are equivalent:

i) φ is flat in p

ii) The germs of coordinate functions $t_1,..,t_n$ of $\mathbb{C}^n$ at 0 are an $\mathcal{O}_{X,p}$-sequence.

This criterion is very useful as we shall see immediately.

3.17. Proposition. If X and Y are arbitrary complex spaces then the canonical projection

$$\pi: X \times Y \longrightarrow Y$$

is flat.

Proof. We first consider the special case $Y = \mathbb{C}^n$. Then for any $x \in X$ we have a canonical isomorphism (see 0.28)

$$\mathcal{O}_{X \times \mathbb{C}^n, (x,o)} \longrightarrow \mathcal{O}_{X,x}[\langle t_1,..,t_n\rangle] =: S \ .$$

It is obvious that $(t_1,..,t_n)$ is an S-sequence. Hence 3.16 implies the assertion.
In the general case we may assume Y is a closed complex subspace of some open subset $Z \subset \mathbb{C}^n$. Then we have a cartesian square

$$\begin{array}{ccc} X \times Y & \longrightarrow & X \times Z \\ \downarrow & & \downarrow \\ Y & \longrightarrow & Z \end{array}$$

where the vertical arrows are the canonical projections. But this reduces the general case to the special case via change of base (3.15).

Corollary. Every locally trivial holomorphic map is flat.

3.18. We state two results of J. FRISCH [35], [36].

Theorem. Let $\varphi: X \to Y$ be a holomorphic map and F a coherent $\mathcal{O}_X$-module. Then the *non-flat locus*

$$\{p \in X: F \text{ is not } \varphi\text{-flat in } p\}$$

is an analytic subset of X.

R. KIEHL [ ] has shown how this result can be derived from the analogous result in algebraic geometry.

Now we consider a fixed holomorphic map $\varphi: X \to Y$ and a coherent $\mathcal{O}_X$-module F. For an arbitrary holomorphic map $\sigma: S \to Y$ we consider the cartesian square

$$\begin{array}{ccc} X_S & \xrightarrow{\ \overline{\sigma}\ } & X \\ {\scriptstyle \varphi_S}\downarrow & & \downarrow{\scriptstyle \varphi} \\ S & \xrightarrow{\ \sigma\ } & Y \end{array}$$

and the coherent $\mathcal{O}_{X_S}$-module $F_S := \overline{\sigma}^*F$. With these notations in mind we formulate the

Flatification Theorem. Assume $\varphi: X \to Y$ is a proper holomorphic map and F is a coherent $\mathcal{O}_X$-module. Then there is a holomorphic map $\tau: T \to Y$ with the following properties:

1) τ is a bijective immersion
2) F_T is φ_T-flat
3) For every holomorphic map $\sigma: S \to Y$ such that F_S is φ_S-flat there exists a unique holomorphic map $\xi: S \to T$ such that $\sigma = \tau \circ \xi$.

T is obtained by cutting Y into pieces according to the position and structure of the non-flat loci.

An alternative method of making F flat is due to HIRONAKA. There it is required that the map σ be a modification. For details we refer to [67].

3.19. Proposition. Every flat holomorphic map is open.

This result was first obtained by DOUADY [27]. The proof given here is due to KIEHL [85]. We need the following

Lemma. If $\varphi: X \to Y$ is flat then for every $p \in X$ the dimension formula

$$\dim_p X = \dim_{\varphi(p)} Y + \dim_p X_{\varphi(p)}$$

holds.

Proof. By 3.15 we may assume Y is reduced. We proceed by induction on $n := \dim_{\varphi(p)} Y$.

The case $n = 0$ is obvious. For $n > 0$ we may find a non-zerodivisor $b \in \mathcal{O}_{Y,\varphi(p)}$. The lemma of 3.12 implies that $a := \tilde{\varphi}_p(b) \in \mathcal{O}_{X,p}$ is a non-zerodivisor. The question being local we may assume we have a cartesian square

$$\begin{array}{ccc} X' & \longrightarrow & X \\ {\scriptstyle \varphi'}\downarrow & & \downarrow{\scriptstyle \varphi} \\ Y' & \longrightarrow & Y \end{array}$$

where Y' resp. X' are the subspaces defined by b resp. a. Since

$$\dim_{\varphi(p)} Y' = \dim_{\varphi(p)} Y - 1$$

and φ' is again flat the induction hypothesis implies the dimension formula for φ'. Since

$$\dim_p X' = \dim_p X - 1$$

the assertion follows.

Proof of the proposition. If $\varphi: X \longrightarrow Y$ is flat and Y is locally irreducible, then φ is open by the above lemma together with Corollary 1 of 3.9.

In the general case we may assume Y is reduced and the normalization map $\nu: \tilde{Y} \longrightarrow Y$ yields a cartesian square

$$\begin{array}{ccc} X \times_Y \tilde{Y} & \longrightarrow & X \\ {\scriptstyle\varphi'}\downarrow & & \downarrow{\scriptstyle\varphi} \\ \tilde{Y} & \longrightarrow & Y \end{array} .$$

φ' is flat by 3.15, hence open as we have seen above. Since Y has the quotient topology with respect to ν this implies φ is open.

3.20. Of course an open holomorphic map need not be flat (use 3.13 to construct obvious examples). But under very auspicious circumstances the miracle may happen (see [80], [82]).

Lemma. Let $f: X \longrightarrow \mathbb{C}$ be an open holomorphic map, take $p \in X$ and assume $f(p) = 0$. If the fibre X_o is reduced then $f_p \in \mathcal{O}_{X,p}$ is a non-zerodivisor and X is reduced in p.

Proof. Put $a := f_p$ and denote by $\mathfrak{a} \subset \mathcal{O}_{X,p}$ the ideal generated by a. If $b \in \mathcal{O}_{X,p}$ and $a \cdot b = 0$ we want to prove $b = 0$.

Take a representative $g \in \mathcal{O}_X(U)$ of b, where $U \subset X$ is some open neighbourhood of p. If $q \in X_o \cap U$ and $g(q) \neq 0$ then $f(x) = 0$ for all x of a neighbourhood $V \subset X$ of q. But then f would not be open. This implies $g(q) = 0$ and

$$b \in \operatorname{rad} \mathfrak{a} = \mathfrak{a}, \text{ i.e. } b = b_1 \cdot a$$

for some $b_1 \in \mathcal{O}_{X,p}$. Since $a^2 \cdot b_1 = 0$ we conclude

$$b \in \mathfrak{a}^2$$

since f^2 is again open. Finally we obtain

$$b \in \bigcap_{n \geq 1} \mathfrak{a}^n = (0) .$$

Now we show that $\mathcal{O}_{X,p}$ is reduced. Take

$$c \in \mathcal{O}_{X,p} \quad \text{with} \quad c^k = 0 .$$

Obviously we have

$$c \in \operatorname{rad} \mathfrak{a} = \mathfrak{a}, \text{ i.e. } c = c_1 \cdot a$$

for some $c_1 \in \mathcal{O}_{X,p}$. Since a is a non-zerodivisor we conclude

$$c_1^k = 0 \quad \text{and} \quad c \in \mathfrak{a}^2 .$$

As above iteration implies $c = 0$.

We recall a well known algebraic fact (see [157]):

Remark. If (R,m) is a Cohen-Macaulay ring and $a \in m$ is an active element, then a is a non-zerodivisor and R/aR is again Cohen-Macaulay.

Proposition. Let $\varphi: X \to \mathbb{C}^n$ be an open holomorphic map and fix $p \in X$.

a) φ reduced in $p \Rightarrow X$ reduced in p and φ flat in p.

b) $\mathcal{O}_{X,p}$ Cohen-Macaulay $\Rightarrow \varphi$ flat in p.

Proof [80], [82]. We proceed by induction on n. The case $n = 0$ is obvious since over $\mathbb{C}$ every module is flat.
For $n > 0$ we may assume $\varphi(p) = 0$. Put $\varphi =: (f_1,..,f_n)$ and consider the cartesian square

$$\begin{array}{ccc} X' & \longrightarrow & X \\ {\scriptstyle \varphi'}\downarrow & & \downarrow{\scriptstyle \varphi} \\ \mathbb{C}^{n-1} & \longrightarrow & \mathbb{C}^n \end{array} ,$$

where $\mathbb{C}^{n-1}$ resp. X' is the subspace defined by the coordinate function z_n resp. f_n .
In case a) φ' is open and reduced in p, hence by induction hypothesis φ' is flat in p and X' is reduced in p. Since X' is the fibre of the open map

$$f_n: X \longrightarrow \mathbb{C}$$

the lemma implies that the germ of f_n in $\mathcal{O}_{X,p}$ is a non-zerodivisor. By standard arguments this implies

$$\mathrm{Tor}_1^R(\mathcal{O}_{X,p}, \mathbb{C}[\langle z_1,..,z_n\rangle]/(z_n)) = 0 .$$

Now 3.14 implies that φ is flat in p and by the above lemma X is reduced in p .
In case b) we use Lemma 1 of 3.10 to conclude that the germ of f_n is active in $\mathcal{O}_{X,p}$. By using the above remark the induction works.

For complex manifolds this proposition together with the results of 3.10 yields the important

Corollary. For a holomorphic map $\varphi: X \to Y$ between connected complex manifolds the following conditions are equivalent:

i) φ is flat.

ii) φ is open.

iii) Every fibre of φ is of pure dimension $\dim X - \dim Y$.

3.21. Another "continuity property" of flat maps is the following regularity criterion (see [52]).

Theorem. If $\varphi: X \to Y$ is a holomorphic map then for $p \in X$ the following conditions are equivalent:

i) φ is a submersion at p (see 2.18)

ii) φ is flat in p and the fibre $X_{\varphi(p)}$ is a manifold at p.

Proof. i) $\Rightarrow$ ii) follows immediately from 3.17. In order to prove ii) $\Rightarrow$ i) we put

$$R := \mathcal{O}_{Y,\varphi(p)}\ ,\quad S := \mathcal{O}_{X,p}\quad\text{and}\quad k := \dim_p X_{\varphi(p)}\ .$$

According to 0.35 we may find an immersion of φ at p, i.e. a commutative diagram

$$\begin{array}{ccc} S & \xleftarrow{\ \alpha\ } & R[\langle t_1,..,t_k\rangle] =: R' \\ & \nwarrow & \uparrow \\ & & R\ , \end{array}$$

where α is surjective and the homomorphism

$$S/mS \xleftarrow{\ \alpha_o\ } R'/mR' \;\cong\; \mathbb{C}[\langle t_1,..,t_k\rangle]$$

induced on the fibres is an isomorphism. Assertion i) is clear if we can prove α is an isomorphism.

Denote by $m \subset R$ the maximal ideal and consider the following diagram:

$$\begin{array}{ccc} m^nS/m^{n+1}S & \xleftarrow{\qquad \alpha_n \qquad} & m^nR'/m^{n+1}R' \\ \downarrow\beta & & \downarrow\beta' \\ (m^n/m^{n+1})\otimes_{R/m}S/mS & \xleftarrow{\quad 1\otimes\alpha_o \quad} & (m^n/m^{n+1})\otimes_{R/m}(R/m)[\langle t_1,..,t_k\rangle] \end{array}$$

for $n \geq 0$, where α_n is the homomorphism induced by α. By assumption S is R-flat and from 3.17 we know R' is R-flat. We claim that these facts yield isomorphisms β and β'.

By tensoring the exact sequence of R-modules

$$0 \longrightarrow m^n \longrightarrow R \longrightarrow R/m^n \longrightarrow 0$$

with S we obtain an isomorphism

$$m^n\otimes_R S \;\longrightarrow\; m^nS\ . \tag{*}$$

If we form the tensor product with S of

$$0 \longrightarrow m^{n+1} \longrightarrow m^n \longrightarrow m^n/m^{n+1} \longrightarrow 0$$

then together with (*) we obtain isomorphisms

$$m^nS/m^{n+1}S \longrightarrow (m^n/m^{n+1})\otimes_R S . \qquad (**)$$

Since this module is annihilated by m we have an isomorphism

$$(m^n/m^{n+1})\otimes_R S \longrightarrow (m^n/m^{n+1})\otimes_{R/m} S/mS . \qquad (***)$$

Now the composition of (**) and (***) yields β. In the same way we obtain β' and the given diagram commutes.
As we know α_o was injective. Since R/m is a field $1\otimes\alpha_o$ is again injective, hence α_n is injective for every $n \geq 0$. But then KRULLs intersection theorem implies that α is injective and the assertion follows.

3.22. For technical reasons it is often useful to know that certain sets of bad behaviour of a holomorphic map φ: $X \to Y$ are analytic. We consider the following subsets of X:

$$S(\varphi) := \{p \in X: X_{\varphi(p)} \text{ is not a manifold in } p\},$$
$$N(\varphi) := \{p \in X: X_{\varphi(p)} \text{ is not normal in } p\},$$
$$R(\varphi) := \{p \in X: X_{\varphi(p)} \text{ is not reduced in } p\}.$$

Proposition. If φ: $X \to Y$ is flat, then the subsets

$$S(\varphi),\ N(\varphi),\ R(\varphi) \subset X$$

are analytic.

We use the following

Lemma. If φ: $X \to Y$ is a holomorphic map then for any $k \in \mathbb{N}$ the sets

$$S_k(\mathcal{O}_X) := \{p \in X: \text{prof } \mathcal{O}_{X,p} \leq k\} \quad \text{and}$$
$$S_k(\mathcal{O}_X;\varphi) := \{p \in X: \text{prof } (\mathcal{O}_{X,p}/m_{Y,\varphi(p)}\mathcal{O}_{X,p}) \leq k\}$$

are analytic in X.

For a proof of the lemma we refer to [137] and [6].

Proof of the proposition (due to C. BĂNICĂ. See also [84] and [6]). We first prove that $S(\varphi) \subset X$ is analytic. For this purpose we use the linear space T(X/Y) over X (see 2.7). It follows from the remark in 2.7 and from 2.3 that

$$X_{\varphi(p)} \text{ manifold at } p \iff \dim_p X_{\varphi(p)} = \dim_{\mathbb{C}} T_p(X/Y) .$$

This immediately implies

$$S(\varphi) = \bigcup_{k \in \mathbb{N}} (\{p \in X: \dim_{\mathbb{C}} T_p(X/Y) \geq k\} \cap \{p \in X: \dim_p X_{\varphi(p)} < k\}) .$$

The first set in the bracket is analytic by the lemma of 2.13 and the second set is the complement of an analytic set by 3.6. Since the union is locally finite, $S(\varphi)$ is locally constructible. But since φ is flat, 3.21 implies that $S(\varphi)$ is closed. Hence $S(\varphi)$ is analytic by 0.46.

Now let us prove $R(\varphi) \subset X$ is analytic. By 2.16 an arbitrary complex space Z is reduced if and only if for every open $U \subset Z$ the restriction map

$$\mathcal{O}_Z(U) \longrightarrow \mathcal{O}_Z(U \setminus S(Z))$$

is injective. By use of local cohomology (see [137], 1.14) we obtain

$$Z \text{ reduced in } z \iff \dim_z(S(Z) \cap S_k(\mathcal{O}_Z)) \leq k-1 \quad \text{for } k \geq 0.$$

This immediately implies

$$R(\varphi) = \bigcup_{k \in \mathbb{N}} \{p \in X : \dim_p(S(\varphi) \cap S_k(\mathcal{O}_X;\varphi) \cap X_{\varphi(p)}) > k-1\} .$$

For fixed k a set of this locally finite union is a degeneracy set of the holomorphic map

$$S(\varphi) \cap S_k(\mathcal{O}_X;\varphi) \longrightarrow Y$$

which is the restriction of φ, hence analytic by 3.6. This proves that $R(\varphi)$ is analytic.

The proof for $N(\varphi)$ is analogous and left to the reader (use Theorem 2.27).

★ ★ ★

CHAPTER 4. <u>MODIFICATIONS AND MEROMORPHIC FUNCTIONS</u>

In this last chapter we want to prove some basic facts about meromorphic functions. In several variables they may have points of indeterminacy; but it turns out that these indeterminacies can be eliminated by a modification of the given space. The main purpose of this chapter is to derive some classical results on the existence of meromorphic functions.

<u>4.1.</u> The most important modification of a complex space is obtained by "blowing up" a certain subspace. In algebraic geometry this procedure is called a monoidal transformation; in complex analysis this method was introduced by HOPF as "σ-Prozess" (see [71]). The technique of blowing up arbitrary subspaces has been developed by HIRONAKA (see [68], [28]).

<u>Theorem.</u> Let X be a complex space with a closed complex subspace $Z \hookrightarrow X$. Then there exists a holomorphic map

$$\sigma: X^* \longrightarrow X$$

with the following properties:

a) σ is proper

b) $Z^* := \sigma^{-1}(Z) \hookrightarrow X^*$ is a hypersurface (see 0.45).

c) σ is universal with respect to b), i.e. if there is any holomorphic map $\tau: X' \to X$ such that $Z' := \tau^{-1}(Z) \hookrightarrow X'$ is a hypersurface then there is a unique holomorphic map $\varphi: X' \to X^*$ such that the diagram

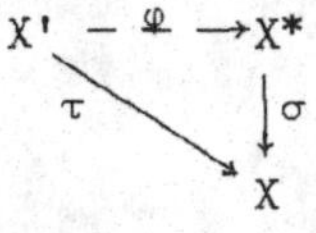

commutes.

d) The restriction of σ

$$X^* \setminus Z^* \longrightarrow X \setminus Z$$

is biholomorphic.

e) If X is a manifold and Z is a submanifold, then X^* is a manifold.

$\sigma: X^* \to X$ is called the *σ-modification* (or *monoidal transformation*) with *center* Z .

<u>Proof.</u> Denote by $J \subset \mathcal{O}_X$ the ideal defining Z and consider the holomorphic map (see 1.9)

$$\xi: \mathbb{P}(J) \longrightarrow X .$$

This is the projective variety over X associated to the linear fibre space

$$\underline{V}(J) \longrightarrow X$$

which is called the *normal space* of Z. Now we define

$$X^* := \mathrm{clos}_{\mathbb{P}(J)}(\,\mathbb{P}(J) \setminus \xi^{-1}(Z)) \quad \text{and} \quad \sigma := \xi|X^*.$$

Geometrically, for every point $p \in Z$ the fibre $\sigma^{-1}(p)$ may be considered as the space of directions normal to Z in which p can be approached from outside Z.
The proof of the stated properties of σ will be given in two steps.

First we consider the special case where $X = \mathbb{C}^n$ with $n \geq 0$ and $Z = (\{0\},\mathbb{C})$ is the simple origin. Then J is generated by the coordinate functions $x_1,..,x_n$.
We use the following well known result which can be verified by an elementary computation.

Lemma. Consider the power series ring $R = \mathbb{C}[\langle x_1,..,x_n\rangle]$ with maximal ideal m. Then the kernel of the epimorphism

$$R^n \longrightarrow m,\ (a_1,..,a_n) \mapsto a_1x_1 + \dots a_nx_n\ ,$$

is generated by the elements

$$(0,..,0,x_j,0,..,0,-x_i,0,..,0) \in R^n$$

where $1 \leq i < j \leq n$ and x_j resp. $-x_i$ is the i-th resp. j-th component.
More generally it can be shown that there is an exact sequence

$$0 \longrightarrow R^{\binom{n}{n}} \rightarrow R^{\binom{n}{n-1}} \rightarrow \dots \rightarrow R^{\binom{n}{k}} \rightarrow \dots \rightarrow R^{\binom{n}{2}} \longrightarrow R^{\binom{n}{1}} \rightarrow m \longrightarrow 0$$

which is called a *chain of syzygies* (see [60], [130]).
According to this lemma there is an exact sequence

$$\mathcal{O}_X^{\binom{n}{2}} \xrightarrow{\ \alpha\ } \mathcal{O}_X^{\,n} \longrightarrow J \longrightarrow 0\ ,$$

where α is given by the holomorphic matrix

$$\begin{bmatrix} x_2 & \vdots & \vdots \\ -x_1 & \vdots & \vdots \\ \dots & x_j \ \dots & \vdots \\ & \vdots & \vdots \\ & -x_i & \vdots \\ & \vdots & x_n \\ & \vdots & -x_{n-1} \end{bmatrix} .$$

If $(x_1,..,x_n,z_1,..,z_n)$ are coordinates in $X \times \mathbb{C}^n$ then

$$V(J) \hookrightarrow X \times \mathbb{C}^n$$

is the linear subspace determined by the linear forms

$$x_j z_i - x_i z_j \ , \quad 1 \le i < j \le n \ ,$$

(see 1.6). The same equations define

$$\mathbb{P}(J) \hookrightarrow X \times \mathbb{P}_{n-1}(\mathbb{C})$$

if we consider $(x_1,..,x_n;z_1:...:z_n)$ as homogeneous coordinates in $X \times \mathbb{P}_{n-1}(\mathbb{C})$.

It is obvious that the canonical projection $\sigma: \mathbb{P}(J) \to X$ is proper. We claim that in our special case ($X = \mathbb{C}^n$ and Z the origin) we even have $X^* = \mathbb{P}(J)$, i.e. $\sigma^{-1}(Z) \subset \mathbb{P}(J)$ is analytically rare (see 0.44). Geometrically this is very easy to see if we consider the holomorphic map

$$\pi: \mathbb{P}(J) \longrightarrow \mathbb{P}_{n-1}(\mathbb{C})$$

induced by the second projection of $X \times \mathbb{P}_{n-1}(\mathbb{C})$ (see Figure 7).

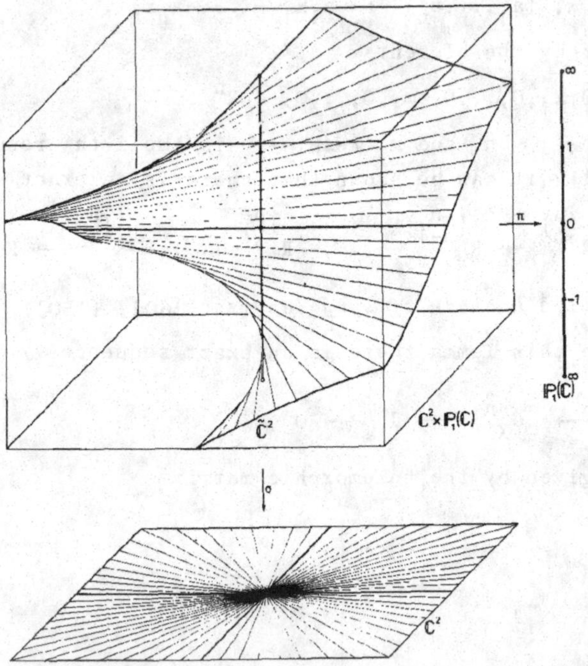

Figure 7

From the equations defining $\mathbb{P}(J)$ one immediately sees that $\mathbb{P}(J)$ is a complex manifold and

$$\mathbb{P}(J) = \{(x,z) \in X \times \mathbb{P}_{n-1}(\mathbb{C}) : x \in z\} ,$$

where $x \in z$ means that $x = (x_1,..,x_n)$ is contained in the line $(z_1:\ldots:z_n) = \mathbb{C}\cdot(z_1,..,z_n)$. Hence for a fixed point $z \in \mathbb{P}_{n-1}(\mathbb{C})$ the fibre $\pi^{-1}(z)$ is a complex line and an elementary computation shows that

$$\pi : \mathbb{P}(J) \longrightarrow \mathbb{P}_{n-1}(\mathbb{C})$$

is a complex line bundle (which is called the HOPF-*bundle*). The zero section of π is just the fibre $\sigma^{-1}(Z)$; this is obviously a hypersurface.

The same fact may be proved in a purely algebraic manner. Denote by $I \subset \mathcal{O}_{X\times\mathbb{P}_{n-1}(\mathbb{C})}$ the sheaf of ideals defining $\mathbb{P}(J)$. For

$$p = (0,..,0;z_1:\ldots:z_n) \in \sigma^{-1}(Z)$$

let us assume $z_1 = 1$. Then the germs at p of

$$x_2 - x_1z_2,\ x_3 - x_1z_3,\ \ldots,\ x_n - x_1z_n \qquad (*)$$

are contained in I_p. Now $\sigma^{-1}(Z) \hookrightarrow \mathbb{P}(J)$ is the subspace generated by the residue classes of $x_1,..,x_n$. Because of the relations (*) the residue class of x_1 alone is sufficient. Since we are in a manifold it is a non-zerodivisor. This proves b) and $X^* = \mathbb{P}(J)$ in the special case. The proof of d) is straightforward and left to the reader.

Next we prove the universal property c) in the special case. It follows from d) that φ is uniquely determined on $X' \setminus Z'$. Since Z' is analytically rare it is sufficient to prove that locally there exists an extension of φ to Z'.

Take $p \in Z'$ and denote by $J \subset \mathcal{O}_{X'}$ the ideal of $Z' \hookrightarrow X'$. By definition J is generated by the functions

$$x_1\circ\tau,\ldots,x_n\circ\tau .$$

On the other hand there is an open neighbourhood $U \subset X'$ of p and a non-zerodivisor $h \in \mathcal{O}_{X'}(U)$ generating J on U. This implies that there are $g_1,..,g_n \in \mathcal{O}_{X'}(U)$ without common zeroes such that

$$x_1\circ\tau = g_1\cdot h ,\ldots,\ x_n\circ\tau = g_n\cdot h \quad \text{on } U$$

if U is sufficiently small. Consider the holomorphic map

$$g = (g_1,..,g_n): U \longrightarrow \mathbb{C}^n \setminus \{0\}$$

and the canonical map

$$\rho: \mathbb{C}^n \setminus \{0\} \longrightarrow \mathbb{P}_{n-1}(\mathbb{C}),\ (z_1,..,z_n) \longmapsto (z_1:\ldots:z_n) .$$

Then the holomorphic map

$$(\tau,\rho\circ g)\colon U \longrightarrow X \times \mathbb{P}_{n-1}(\mathbb{C})$$

has its image in X^* and extends φ as one immediately verifies.

Now we have to attack the general case. It is obvious from the definition that our given map

$$\sigma\colon X^* \longrightarrow X$$

is proper. The proof of the remaining properties is a problem which is local with respect to X. Hence we may assume there is a holomorphic map

$$\varphi\colon X \longrightarrow \mathbb{C}^n \quad \text{with } Z = \varphi^{-1}(0)\,,$$

where $0 \hookrightarrow \mathbb{C}^n$ denotes the origin. We are going to build up the following commutative diagram of holomorphic maps:

$$\begin{array}{ccccc}
Z^* & \longleftrightarrow & Z_1 & \longrightarrow & 0^* \\
\downarrow & & \downarrow & & \downarrow \\
X^* & \longleftrightarrow & X_1 & \longrightarrow & (\mathbb{C}^n)^* \\
& \searrow & \downarrow & & \downarrow \\
& & X & \xrightarrow{\ \varphi\ } & \mathbb{C}^n \\
& & \uparrow & & \uparrow \\
& & Z & \longrightarrow & 0\,.
\end{array}$$

Let us now denote by $m \subset \mathcal{O}_{\mathbb{C}^n}$ the sheaf of ideals of the origin 0. Then we put

$$(\mathbb{C}^n)^* := \mathbb{P}(m) \quad \text{and} \quad X_1 := \mathbb{P}(\varphi^* m)\ .$$

Together with the canonical projection

$$(\mathbb{C}^n)^* \longrightarrow \mathbb{C}^n$$

we obtain the σ-modification with the origin as center as we have seen above. By 1.8 we know

$$X_1 = X \times_{\mathbb{C}^n} (\mathbb{C}^n)^*.$$

To give a more explicit description we use the presentation of m from above. It yields an exact sequence

$$\mathcal{O}_X^{\binom{n}{2}} \longrightarrow \mathcal{O}_X^{\,n} \longrightarrow \varphi^* m \longrightarrow 0$$

and if $\varphi = (f_1,\ldots,f_n)$ we obtain

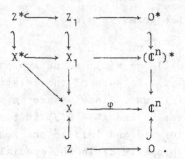

$$\mathbb{P}(\varphi^* m) \hookrightarrow X \times \mathbb{P}_{n-1}(\mathbb{C})$$

as the closed complex subspace defined by the forms

$$f_j z_i - f_i z_j \ , \quad 1 \le i < j \le n \ .$$

It is easy to see that the inverse image $Z_1 \hookrightarrow X_1$ of Z is defined by a sheaf of ideals which is principal (see 0.45; take the residue class of f_i if $z_i \neq 0$).

However, in general, Z_1 is not a hypersurface (see Figure 8). In 0.45 we have seen what has to be done.

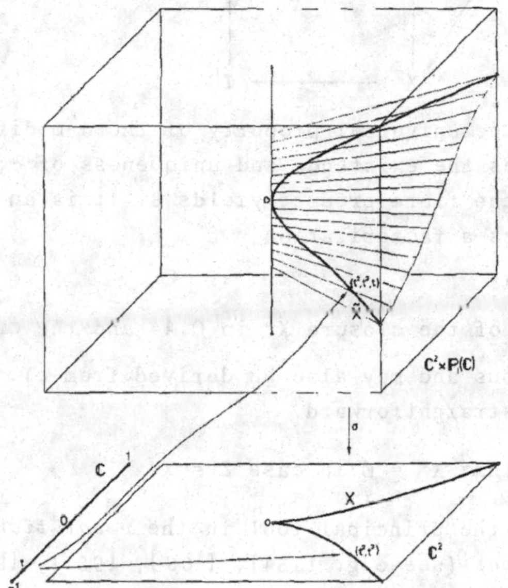

Figure 8

Let us first remark that there is a canonical epimorphism

$$\varphi^* m \longrightarrow I \ ,$$

where $I \subset \mathcal{O}_X$ is the sheaf of ideals defining Z (see 0.27). Hence we know that

$$\mathbb{P}(I) \hookrightarrow \mathbb{P}(\varphi^* m) = X_1$$

is a closed complex subspace. Outside the inverse image of Z they coincide. This implies

$$X^* := \operatorname{clos}_{\mathbb{P}(I)}(\mathbb{P}(I) \smallsetminus \xi^{-1}(Z)) = \operatorname{clos}_{X_1}(X_1 \smallsetminus Z_1) \ .$$

But now by 0.45 we know $Z^* \hookrightarrow X^*$ is a hypersurface.

What happens geometrically can also be seen from Figure 8 : the components of X_1 lying entirely in Z_1 have to be removed. In the example of the cusp X , the space X^* is a smooth curve an 0^* is a double point.

Now the universal property c) is easy to verify. Given a competitor $\tau: X' \to X$ we catch him in the following diagram:

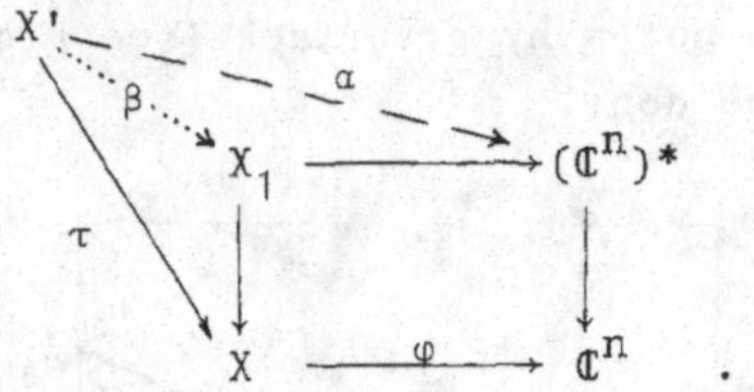

By considering $\varphi\circ\tau$ the universal property of the σ-modification of $\mathbb{C}^n$ with center 0 implies the existence and uniqueness of α; then the universal property of the fibre product yields β. It is an easy exercise to show that β admits a factorization

$$X' \xrightarrow{\varphi} X^* \hookrightarrow X_1$$

(use the definition of the closure X^* in 0.44 and the criterion of 0.17).

Property d) is obvious and may also be derived from c).
The proof of e) is straightforward.

It should be noted that $X^* = \emptyset$ in case $Z = X$.

σ-modifications are the principal tool in the *resolution of singularities* of complex spaces (see e.g. [154], [69], [95], [96], [61], [ ]).
A very simple example is the σ-modification of the cusp in the origin (Fig. 8). Blowing up an arbitrary subspace of a manifold may yield singularities.

Example. Consider $\mathbb{C}^2$ with coordinates x_1, x_2 and take as center $Z \hookrightarrow \mathbb{C}^2$ the multiple point defined by $x_1{}^2 = x_2{}^2 = 0$. Then

$$X := (\mathbb{C}^2)^* = \{(x_1,x_2;z_1:z_2) \in \mathbb{C}^2 \times \mathbb{P}_1(\mathbb{C}):\ x_2{}^2 z_1 - x_1{}^2 z_2 = 0\}$$

(Figure 9). This space has singularities along $\{0\} \times \mathbb{P}_1(\mathbb{C})$. This is the non-normal locus $N(X)$. The normalization may be given by

$$\tilde{X} := \{(x_1,x_2;z_1:z_2) \in \mathbb{C}^2 \times \mathbb{P}_1(\mathbb{C}):\ x_2 z_1 - x_1 z_2 = 0\}\ .$$

This is the σ-modification of $\mathbb{C}^2$ with the simple origin as center.

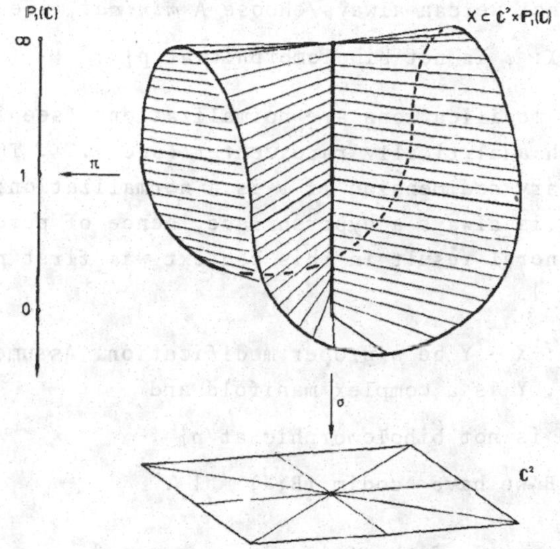

Figure 9

The normalization map

$$\nu\colon \tilde{X} \longrightarrow X$$

is just the restriction of the holomorphic map

$$\mathbb{C}^2 \times \mathbb{P}_1(\mathbb{C}) \longrightarrow \mathbb{C}^2 \times \mathbb{P}_1(\mathbb{C}) , \quad (x_1,x_2;z_1:z_2) \longmapsto (x_1,x_2;z_1^2:z_2^2) .$$

Over the non-normal locus we obtain the covering

$$\mathbb{P}_1(\mathbb{C}) \longrightarrow \mathbb{P}_1(\mathbb{C}), \ (z_1:z_2) \longmapsto (z_1^2:z_2^2) ,$$

which obviously does not admit a continuous section (compare 2.28).

4.2. In the case of a σ-modification points are always replaced by projective varieties. In complex analysis it is necessary to consider modifications with arbitrary fibres. We give the following definition (see [46], [93]).

Let $\varphi\colon X \to Y$ be a holomorphic map between complex spaces. φ is called a *proper modification* if it is proper and if there is an analytic set $A \subset Y$ with the following properties:

1) $A \subset Y$ and $\varphi^{-1}(A) \subset X$ are analytically rare
2) the restriction of φ

   $$X \setminus \varphi^{-1}(A) \longrightarrow Y \setminus A$$

   is biholomorphic.

It is easy to see that we can always choose A *minimal*, i.e. such that

$\varphi^{-1}(A) = \{p \in X: \varphi$ is not biholomorphic at $p\}$.

Examples for proper modifications are normalizations (see 2.26) or σ-modifications with analytically rare center (see 4.1). The set $\varphi^{-1}(A) \subset X$ may have arbitrary codimension if φ is a normalization; in case of a σ-modification it is always a hypersurface, hence of pure codimension 1. The following general result in this context was first proved by GRAUERT-REMMERT [46].

Proposition. Let $\varphi: X \to Y$ be a proper modification. Assume X is a reduced complex space, Y is a complex manifold and

$B := \{p \in X: \varphi$ is not biholomorphic at $p\}$.

Then for every $p \in B$ we have $\operatorname{codim}_p(B;X) = 1$.

It is an easy consequence of the following

Purity of branch theorem. Let $\varphi: X \to Y$ be a holomorphic map and consider the analytic set

$B := \{p \in X: \varphi$ is not biholomorphic at $p\}$.

Assume X is normal, Y is a manifold and B is analytically rare. Then for any $p \in B$ we have $\operatorname{codim}_p(B;X) = 1$.

This result is due to GRAUERT-REMMERT [46]. The analogous assertion in algebraic geometry was proved by ZARISKI [155]. A short complex analytic proof using hulls of holomorphy was given by KERNER [81]. For the algebraic background we refer to KUHLMANN [98].

We show how to derive the above proposition from this theorem (see [81]).

If X is normal this is clear. Otherwise we use the normalization

$\nu: \tilde{X} \longrightarrow X$.

Then $\tilde{\varphi} := \varphi \circ \nu$ is again a proper modification. Consider the sets

$B = \{p \in X: \varphi$ is not biholomorphic at $p\}$ and

$\tilde{B} = \{q \in X: \tilde{\varphi}$ is not biholomorphic at $q\}$.

Now it is easy to check that $\nu(\tilde{B}) = B$ and this implies the assertion in general.

It should be noted that the purity of branch theorem is obvious in case X is a manifold (see 2.17).

The so called Lemma of CHOW states that an arbitrary proper modification can be "dominated" by a σ-modification. For details we refer to [63] and [103].

4.3. In 1949 CHOW [25] proved that every analytic subset of a complex projective space is algebraic. Simplified proofs were given by H. KNESER [88] and REMMERT-STEIN [118]. In 1955 SERRE [129] showed that the theorem of CHOW is a special case of a general correspondence theorem between algebraic and analytic geometry. It is often useful to have a "relative" (i.e. depending on a parameter) version of these theorems. This was established by HAKIM [58] (see also [47], [32]).

CHOWs Theorem. Let X be an arbitrary complex space. Then every closed complex subspace

$$Y \hookrightarrow X \times \mathbb{P}_n(\mathbb{C})$$

is *relatively algebraic*, i.e. there is a cone $K \hookrightarrow X \times \mathbb{C}^{n+1}$ over X such that (see 1.3) $Y = \mathbb{P}(K)$.

Proof. We use the σ modification

$$\sigma\colon (\mathbb{C}^{n+1})^* \longrightarrow \mathbb{C}^{n+1}$$

with the origin as center (see 4.1). By definition $(\mathbb{C}^{n+1})^*$ is a closed complex subspace of $\mathbb{C}^{n+1} \times \mathbb{P}_n(\mathbb{C})$. The restriction of the second projection yields a complex line bundle

$$\pi\colon (\mathbb{C}^{n+1})^* \longrightarrow \mathbb{P}_n(\mathbb{C}) .$$

Finally we denote by

$$\rho\colon \mathbb{C}^{n+1} \setminus \{0\} \longrightarrow \mathbb{P}_n(\mathbb{C}) , \ (z_o,\ldots,z_n) \longmapsto (z_o:\ldots:z_n) ,$$

the canonical map. If we define

$$\sigma' := id_X \times \sigma, \ \pi' = id_X \times \pi \ \text{ and } \ \rho' = id_X \times \rho$$

we obtain the commutative diagram

$$\begin{array}{ccccc} & & X \times (\mathbb{C}^{n+1})^* & & \\ & \swarrow^{\sigma'} & & \searrow^{\pi'} & \\ X \times \mathbb{C}^{n+1} & & & & X \times \mathbb{P}_n(\mathbb{C}) \\ \cup & & & \nearrow_{\rho'} & \\ X \times (\mathbb{C}^{n+1} \setminus \{0\}) & & & & \end{array} .$$

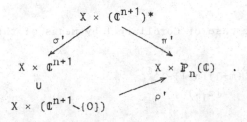

The subspace $\rho'^{-1}(Y) \longrightarrow X \times (\mathbb{C}^{n+1} \setminus \{0\})$ consists of cones without peaks. The problem is to find an analytic extension to cones with peaks. The above diagram suggests how this can be done.
σ' is proper since σ is proper, hence we obtain a closed complex subspace

$$K := \sigma'(\pi'^{-1}(Y)) \longrightarrow X \times \mathbb{C}^{n+1} .$$

We show that K is a cone over X. If

$$\mu: \mathbb{C} \times X \times \mathbb{C}^{n+1} \longrightarrow X \times \mathbb{C}^{n+1}$$

denotes the relative scalar multiplication we have to prove

$$\mathbb{C} \times K \longrightarrow \mu^{-1}(K) \qquad (*)$$

is a closed complex subspace (see 1.2). Since

$$K \cap (X \times (\mathbb{C}^{n+1} \setminus \{0\})) = \rho'^{-1}(Y) \qquad (**)$$

we know that (*) holds outside $\mathbb{C} \times X \times \{0\}$. Since this set is analytically rare in $\mathbb{C} \times K$, (*) holds everywhere.
Finally (**) implies $Y = \mathbb{P}(K)$.

In case X is a simple point we obtain the

Corollary 1. Every closed complex subspace of $\mathbb{P}_n(\mathbb{C})$ is an algebraic variety.

It should be noted that an analogous result in the real case does not hold.

Example. Put $U := \{(x_1,x_2) \in \mathbb{R}^2: -\frac{\pi}{2} < x_1 < \frac{\pi}{2}\}$ and

$$Y = \{(x_1,x_2) \in U: \tan^2 x_1 + x_2^2 = 1\} .$$

$Y \subset \mathbb{R}^2$ is a compact analytic subset, hence Y can be considered as an analytic curve without points at infinity in $\mathbb{P}_2(\mathbb{R})$. With some computation one sees that Y cannot be algebraic.

We finally consider the special case $X = \mathbb{P}_n(\mathbb{C})$, i.e. a closed complex subspace

$$Y \longrightarrow \mathbb{P}_m(\mathbb{C}) \times \mathbb{P}_n(\mathbb{C}) .$$

This may be reduced to the case of Corollary 1 by means of the *Segre embedding*

$$\mathbb{P}_m(\mathbb{C}) \times \mathbb{P}_n(\mathbb{C}) \longrightarrow \mathbb{P}_{(m+1)(n+1)-1}(\mathbb{C})$$
$$(x_i;y_j) \longmapsto (z_{ij}) ,$$

where $0 \leq i \leq m$, $0 \leq j \leq n$ and $z_{ij} = x_i y_j$. Its image is defined by the equations

$$z_{ij}z_{kl} - z_{kj}z_{il} = 0 \quad (0 \leq i,k \leq m \; ; \; 0 \leq j,l \leq n)$$

(for details see e.g. [132]).

Then Y is a closed complex subspace of $\mathbb{P}_N(\mathbb{C})$ where $N = (m+1)(n+1)-1$; hence the ideal of Y is generated by a finite number of homogeneous polynomials in the N+1 variables z_{ij}. By substituting $z_{ij} = x_i y_j$ and with a little bit of computation (see e.g. [132], I §5) one obtains

Corollary 2. Given a closed complex subspace

$$Y \hookrightarrow \mathbb{P}_m(\mathbb{C}) \times \mathbb{P}_n(\mathbb{C})$$

there are polynomials

$$F_1,..,F_k \in \mathbb{C}[x_o,..,x_m;y_o,..,y_n]$$

which are homogeneous with respect to the variables x_i and y_j separately (not necessarily of the same degree) and generate the ideal of Y.

The generalization to an arbitrary finite product of projective spaces is obvious.

4.4. The definition of meromorphic functions on an arbitrary complex space X presents some technical difficulties. Locally every meromorphic function is a quotient of holomorphic functions, where the denominator has to be a non-zerodivisor. Of course this does not yield "functions" in the elementary sense.

The precise definition is as follows: We consider the subsheaf

$S_X \subset \mathcal{O}_X$ defined by $S_{X,p} = \{a \in \mathcal{O}_{X,p} : \text{a is non-zerodivisor}\}$

for all $p \in X$. Given $f \in \mathcal{O}_X(X)$ such that $f_p \in S_{X,p}$ for some $p \in X$ then for some open neighbourhood $U \subset X$ of p the function f is a non-zerodivisor in $\mathcal{O}_X(U)$. This may easily be proved by considering the endomorphism of $\mathcal{O}_X$ defined by multiplication by f and by using 0.6.

Now for any open $U \subset X$ the subset $S_X(U) \subset \mathcal{O}_X(U)$ is multiplicatively closed and contains no zerodivisor. We obtain a presheaf

$$U \longmapsto S_X(U)^{-1}\mathcal{O}_X(U)$$

by associating to U the ring of quotients. However, in general, this is not a sheaf. By

$\mathcal{M}_X$

we denote the associated sheaf. For every $p \in X$ we obtain

$$\mathcal{M}_{X,p} = S_{X,p}^{-1}\mathcal{O}_{X,p} .$$

Now for any open $U \subset X$ a *meromorphic function* f on U is defined as a section $f \in \mathcal{M}_X(U)$. Then for any $p \in U$ we have

$$f_p = \frac{g_p}{h_p}, \text{ with } g_p \in \mathcal{O}_{X,p} \text{ and } h_p \in S_{X,p} .$$

In contrast to a holomorphic function, meromorphic functions may have poles and indeterminacies. We first consider poles.

Since every holomorphic function is meromorphic we obtain a canonical monomorphism

$$\mathcal{O}_X \longrightarrow \mathcal{M}_X$$

(the injectivity results from the fact that S_X contains no zerodivisor). For $f \in \mathcal{M}_X(X)$ we consider the subsheaf

$$\mathcal{O}_X \cdot f \subset \mathcal{M}_X .$$

Then we may define an ideal

$$\mathcal{D} := \mathcal{O}_X : (\mathcal{O}_X \cdot f) \subset \mathcal{O}_X ,$$

which we call the *sheaf of denominators of* f. By definition of $\mathcal{D}$ we have

$$\mathcal{D}_p = \{\beta \in \mathcal{O}_{X,p} : \beta \cdot f_p \in \mathcal{O}_{X,p}\}$$

for every $p \in X$. The closed complex subspace

$$P(f) \hookrightarrow X$$

defined by $\mathcal{D}$ is called the *pole variety of* f. Obviously this is the smallest subset such that f is holomorphic on $X \setminus P(f)$.

In case the local rings $\mathcal{O}_{X,p}$ are factorial for all $p \in X$ the situation is particularly convenient. Then we may represent

$$f_p = \frac{g_p}{h_p} \text{ with } g_p \text{ and } h_p \text{ relatively prime.}$$

If we have another quotient representation with g_p' and h_p' then there are units $\alpha, \beta \in \mathcal{O}_{X,p}$ such that

$$g_p' = \alpha g_p \quad \text{and} \quad h_p' = \beta h_p .$$

This implies that $\mathcal{D}_p \subset \mathcal{O}_{X,p}$ is a principal ideal and $P(f) \hookrightarrow X$ is a hypersurface (see 0.45). This need not be the case in general (see the trivial example below).

Recall that every weakly holomorphic function is meromorphic (this was a consequence of the existence of universal denominators; see 2.24,2.25).

This yields inclusions

$$\mathcal{O}_X \subset \tilde{\mathcal{O}}_X \subset \mathcal{M}_X$$

in case X is reduced. Hence a meromorphic function may be locally bounded even if the "pole variety" is not empty.

Example. In $\mathbb{C}^2$ we consider the reduced complex subspace

$$X := \{(x,y) \in \mathbb{C}^2 : x \cdot y = 0\} .$$

We also denote by $x,y \in \mathcal{O}_X(X)$ the restrictions of the coordinate functions in $\mathbb{C}^2$. Then $x+y$ is a universal denominator at the origin and $\tilde{\mathcal{O}}_X$ is generated as $\mathcal{O}_X$-module by the weakly holomorphic functions

$$\frac{x}{x+y} \quad \text{and} \quad \frac{y}{x+y} .$$

Now consider the meromorphic function

$$f := \frac{x}{x^2+y} \in \mathcal{M}_X(X) .$$

On X we obviously have

$$\frac{x}{x^2+y} = \frac{x}{x^2+y^n}$$

for $n \geq 1$, hence x^2+y^n is a denominator. Another denominator is x^2 since on X

$$\frac{x^3}{x^2+y} = \frac{x}{1} .$$

An easy computation shows that x^2 and y generate the denominator ideal $\mathcal{D}$. The pole variety P(f) is a double point, which is not a hypersurface in X.

It is easy to construct meromorphic functions such that the pole variety even has codimension greater than one. Just take a weakly holomorphic function on a non-normal complex space. However we note the

Remark. If X is a normal complex space and f is meromorphic on X, then the pole variety P(f) is of pure codimension 1.

This is an immediate consequence of the Second RIEMANN Extension Theorem (see 2.27).

We still collect some elementary properties of meromorphic functions.

Remark 2. Let X be a complex space and take a subset $A \subset X$ which is analytically rare. Then for every open $U \subset X$ the restriction homomorphism

$$M_X(U) \longrightarrow M_X(U \setminus A)$$

is injective.

Proof. Take $f \in M_X(U)$ such that $f|(U \setminus A) = 0$. For $p \in U$ write

$$f_p = \frac{g_p}{h_p}.$$

If g_p is represented by $g \in \mathcal{O}_X(V)$ for some open $V \subset U$, then

$$g|(V \setminus A) = 0, \text{ hence } g = 0.$$

This implies $f_p = 0$ for every $p \in U$, hence $f = 0$.

As for holomorphic functions there is an

Identity Theorem. Let X be a reduced and irreducible complex space. If f is meromorphic on X and $f|U = 0$ for some open $U \subset X$, then $f = 0$.

Proof. The assertion is well known in case f is holomorphic (use the fact that the non-singular locus $X \setminus S(X)$ is a connected manifold). If f is meromorphic, then f is holomorphic on the irreducible space $X \setminus P(f)$. Since $P(f)$ is analytically rare we obtain $f = 0$.

For any complex space X, the set $M_X(X)$ is an $\mathcal{O}_X(X)$-module and a $\mathbb{C}$-algebra in a canonical way. In case X is reduced one knows more.

Proposition. Let X be a reduced complex space.

a) If X is irreducible, then $M_X(X)$ is a field.

b) Given the locally finite decomposition into irreducible components

$$X = \bigcup_{i \in I} X_i \text{ , then } M_X(X) = \prod_{i \in I} M_{X_i}(X_i) \text{ ,}$$

i.e. the $\mathbb{C}$-algebra $M_X(X)$ is a direct product of fields.

Proof. a) It suffices to show that every $0 \neq f \in M_X(X)$ has an inverse f^{-1}. For every $p \in X$ we may find an open neighbourhood U with $g, h \in \mathcal{O}_X(U)$ such that

$$f = \frac{g}{h} \text{ on } U.$$

Assume g_p is a zero-divisor in $\mathcal{O}_{X,p}$. Then for sufficiently small U, there is an irreducible component $U' \subset U$ such that $g|U' = 0$. But then the Identity Theorem implies $f = 0$; this yields a contradiction. Choose U so small that g_x is a non-zerodivisor for every $x \in U$. Then we may define

$$f^{-1} := \frac{h}{g} \text{ on } U.$$

Obviously a covering of X with such open sets U yields f^{-1} on X.

b) For every $i \in I$ we define a function

$$\chi_i(x) = \begin{cases} 1 & \text{for } x \in X_i \\ 0 & \text{otherwise.} \end{cases}$$

This function is weakly holomorphic on X, hence meromorphic (use the universal denominators of 2.25).

Now consider the canonical monomorphism

$$M_X(X) \longrightarrow \prod_i M_{X_i}(X_i) \ , \quad f \longmapsto (f|X_i)_{i\in I} \ .$$

Any family $(f_i)_{i\in I}$ of meromorphic functions $f_i \in M_{X_i}(X_i)$ is the image of the locally finite sum

$$f := \sum_{i\in I} \chi_i f_i \ .$$

Hence our map is an isomorphism.

This proposition shows that in case of reduced complex spaces investigations about meromorphic functions may be restricted to the irreducible case without loss of generality.

4.5. In order to study the "points of indeterminacy" of a meromorphic function we construct its "graph". This is a very useful technical device which makes it often easier to handle these complicated objects. We start with an

Example. Take $\mathbb{C}^2$ with coordinates (x,y) and consider the meromorphic function

$$f = \frac{y}{x}$$

in $\mathbb{C}^2$. Outside the line $x = 0$ it is holomorphic and there the graph is given by

$$\{(x,y,z) \in \mathbb{C}^2\times\mathbb{C}: \ x \neq 0 \text{ and } z = \frac{y}{x}\}.$$

If we take the closure of this set in $\mathbb{C}^2 \times \mathbb{P}_1(\mathbb{C})$ we obtain

$$\Gamma_f := \{(x,y;z_o:z_1) \in \mathbb{C}^2 \times \mathbb{P}_1(\mathbb{C}): \frac{z_1}{z_o} = \frac{y}{x}, \text{ i.e. } yz_o - xz_1 = 0\}.$$

The projections of $\mathbb{C}^2 \times \mathbb{P}_1(\mathbb{C})$ on the factors restricted to Γ_f yield holomorphic maps

$$\sigma: \Gamma_f \longrightarrow \mathbb{C}^2 \quad \text{and} \quad \pi: \Gamma_f \longrightarrow \mathbb{P}_1(\mathbb{C}) \ .$$

σ is just the monoidal transformation of $\mathbb{C}^2$ with the origin as center (see 4.1, in particular Figure 7). For every point $p \in \mathbb{C}^2$ we may consider the subset

$$f(p) := \pi(\sigma^{-1}(p)) \subset \mathbb{P}_1(\mathbb{C})$$

as the "value" of f in p. Outside the origin this is a single point; the origin is a point of indeterminacy. If we replace $\mathbb{C}^2$ by Γ_f and f by π then we have an ordinary map and the indeterminacy is eliminated.

Now we consider a general complex space X. If f is a holomorphic function, we obtain a closed complex subspace

$$\Gamma_f \hookrightarrow X \times \mathbb{C}$$

as the fibre product over $\mathbb{C}$ of f with $id_{\mathbb{C}}$. This is called the *graph of the holomorphic map* f. The holomorphic map

$$\sigma: \Gamma_f \longrightarrow X$$

which is the restriction of the first projection is obviously biholomorphic (use the universal property of the fibre product to construct σ^{-1}).

Conversely, if

$$\Gamma \hookrightarrow X \times \mathbb{C}$$

is any closed complex subspace such that the canonical map

$$\sigma: \Gamma \longrightarrow X$$

is biholomorphic, then Γ is the graph of the holomorphic function $\pi\circ\sigma^{-1}$, where

$$\pi: \Gamma \longrightarrow \mathbb{C}$$

is the restriction of the second projection.

For meromorphic functions there are two complications, due to possible poles and indeterminacies.

Definition. Consider a closed complex subspace

$$\Gamma \hookrightarrow X \times \mathbb{P}_1(\mathbb{C}) \quad \text{and denote by} \quad \sigma: \Gamma \longrightarrow X$$

the canonical map. Γ is called a *holomorphic graph at* $p \in X$, if there is an open neighbourhood U of p such that

a) $\sigma|\sigma^{-1}(U): \sigma^{-1}(U) \longrightarrow U$ is biholomorphic

b) $\sigma^{-1}(U) \cap (U\times\{\infty\}) = \emptyset$.

This condition may be tested in the following way. According to the Theorem of CHOW (4.3) there is a sheaf of homogeneous ideals

$$A \subset \mathcal{O}_X[w_0, w_1]$$

associated to Γ. It determines coherent sheaves of *linear coefficient ideals*

$$A^{(o)}, A^{(1)} \subset \mathcal{O}_X$$

with the properties

$$A_p^{(o)} = \{ a \in \mathcal{O}_{X,p}: aw_o+bw_1 \in A_p \text{ for some } b \in \mathcal{O}_{X,p} \} \text{ and}$$

$$A_p^{(1)} = \{ b \in \mathcal{O}_{X,p}: aw_o+bw_1 \in A_p \text{ for some } a \in \mathcal{O}_{X,p} \}.$$

<u>Lemma.</u> With the notations as above, $\Gamma \hookrightarrow X \times \mathbb{P}_1(\mathbb{C})$ is a holomorphic graph in $p \in X$ if and only if the following conditions are satisfied:

α) The canonical homomorphism

$$\tilde{\sigma}_p: \mathcal{O}_{X,p} \longrightarrow (\sigma_*\mathcal{O}_\Gamma)_p$$

is injective.

β) $A_p^{(1)} = \mathcal{O}_{X,p}$.

<u>Proof.</u> We put $R := \mathcal{O}_{X,p}$. The homogeneous ideal $A_p \subset R[w_o,w_1]$ determines ideals

$$\mathfrak{a} \subset R[w] \text{ , where } w = \frac{w_1}{w_o} \text{ , and } \mathfrak{b} \subset R[w'] \text{ , where } w' = \frac{w_o}{w_1} \text{ ,}$$

in the usual way. Consider the homomorphism

$$\varphi: R \longrightarrow R[w]/\mathfrak{a}$$

which is the composite of the canonical homomorphisms

$$R \longrightarrow R[w] \text{ and } R[w] \longrightarrow R[w]/\mathfrak{a} \text{ .}$$

It is easy to check that φ is surjective if and only if

$$a + w \in \mathfrak{a} \text{ for some } a \in R \text{ .}$$

Now assume Γ is a holomorphic graph in p, i.e. it satisfies conditions a) and b) of the definition. It is obvious that a) implies α).
a) and b) together imply that φ is an isomorphism, and as we have seen this yields

$$aw_o + 1w_1 \in A_p \quad \text{for some } a \in R \text{ ,} \qquad (*)$$

hence β).
Conversely β) implies (*) and this yields

$$aw' + 1 \in \mathfrak{b} \text{ ,}$$

hence b). On the other hand (*) implies

$$a + w \in \mathfrak{a} \text{ .}$$

Consequently φ is surjective and $\sigma^{-1}(p)$ consists of a single point q. Furthermore $\tilde{\sigma}_p$ is injective, hence

$$\tilde{\sigma}_p: \mathcal{O}_{X,p} \longrightarrow \mathcal{O}_{\Gamma,q}$$

is an isomorphism and this implies a).

Definition. A closed complex subspace

$$\Gamma \hookrightarrow X \times \mathbb{P}_1(\mathbb{C}) \quad \text{with canonical map } \sigma: \Gamma \longrightarrow X$$

is called a *meromorphic graph over* X, if there is an analytic set $A \subset X$ with the following properties:

1) $A \subset X$ and $\sigma^{-1}(A) \subset \Gamma$ are analytically rare.
2) Γ is a holomorphic graph outside A.

Now we construct a meromorphic graph for every meromorphic function f on a complex space X.
As in 4.4 we denote by

$$\mathcal{D} = \mathcal{O}_X:(\mathcal{O}_X\cdot f) \subset \mathcal{O}_X$$

the denominator ideal of f. Furthermore we define

$$F = \mathcal{O}_X + \mathcal{O}_X\cdot f \subset \mathcal{M}_X ,$$

which is a coherent $\mathcal{O}_X$-module. Then we obtain an exact sequence

$$0 \longrightarrow \mathcal{D} \xrightarrow{\alpha} \mathcal{O}_X{}^2 \xrightarrow{\beta} F \longrightarrow 0 ,$$

where α and β are determined by

$$\alpha_p(a) := (a\cdot f_p, -a) \quad \text{and} \quad \beta_p(b,c) := b + cf_p$$

for every $p \in X$. According to 1.9 we obtain

$$\mathbb{P}(F) \hookrightarrow X \times \mathbb{P}_1(\mathbb{C}) .$$

If we have an open set $U \subset X$ and generators

$$h_1,\ldots,h_r \in \mathcal{D}(U)$$

of $\mathcal{D}$ on U, then over U the homogeneous ideal

$$\mathcal{B} \subset \mathcal{O}_X[w_o,w_1]$$

of $\mathbb{P}(F)$ is generated by

$$h_1 fw_o - h_1 w_1,\ldots, h_r fw_o - h_r w_1 \in \mathcal{O}_X(U)[w_o,w_1] .$$

For the coefficient ideal defined above we have

$$\mathcal{B}^{(1)} = \mathcal{D} .$$

Since $\mathcal{D}_p$ contains a nonzerodivisor for every $p \in X$, the set

$$A := \operatorname{supp}(\mathcal{O}_X/\mathcal{B}^{(1)})$$

is analytically rare. The above lemma implies that $\mathbb{P}(F)$ is a holomorphic graph outside A. But in general $\sigma^{-1}(A) \subset \mathbb{P}(F)$ is not rare (consider the weakly holomorphic function of Example 2 in 2.24). But

$$\Gamma_f := \mathrm{clos}_{X\times\mathbb{P}_1(\mathbb{C})}(\mathbb{P}(F)\smallsetminus\sigma^{-1}(A))$$

satisfies the conditions for a meromorphic graph. If

$$\mathcal{A} \subset \mathcal{O}_X[w_o, w_1]$$

denotes the homogeneous ideal of Γ_f, then $\mathcal{B} \subset \mathcal{A}$ and the linear parts of $\mathcal{A}$ and $\mathcal{B}$ are the same.

<u>4.6.</u> Now we show that conversely every meromorphic graph comes from a meromorphic function (see [28], [150], [93], [32]).

<u>Proposition.</u> For every complex space X the map

$$f \longmapsto \Gamma_f$$

from the set of meromorphic functions on X to the set of meromorphic graphs over X is bijective.

<u>Proof.</u> Given a meromorphic graph

$$\Gamma \hookrightarrow X \times \mathbb{P}_1(\mathbb{C}) \quad \text{with projection } \sigma\colon \Gamma \longrightarrow X$$

we construct a meromorphic function f. The idea is to extract the required numerators and denominators of f from Γ by means of CHOWs Theorem. To do this consider the homogeneous ideal

$$\mathcal{A} \subset \mathcal{O}_X[w_o, w_1]$$

of Γ and the coefficient ideal $\mathcal{A}^{(1)} \subset \mathcal{O}_X$ of w_1 (see 4.5). Now we apply the lemma of 4.5. Since Γ is a holomorphic graph outside an analytically rare set, the map

$$\tilde{\sigma}_p\colon \mathcal{O}_{X,p} \longrightarrow (\sigma_*\mathcal{O}_\Gamma)_p$$

is injective for every $p \in X$. Hence, by the lemma of 4.5, the set of points where Γ is not a holomorphic graph is equal to

$$A := \mathrm{supp}(\mathcal{O}_X/\mathcal{A}^{(1)}) .$$

In particular, $A \subset X$ is analytically rare. Consequently (see 0.43) for every $p \in X$ there is a non zero-divisor

$$h_p \in \mathcal{A}_p^{(1)} .$$

By definition of the coefficient ideals there exists

$$g_p \in \mathcal{O}_{X,p} \quad \text{such that} \quad g_p w_o - h_p w_1 \in \mathcal{A}_p .$$

We want to use g_p and h_p as numerator and denominator of the desired germ $f_p \in M_{X,p}$.
We first show that for given h_p the germ g_p with the above properties is uniquely determined.
Assume we have

$$g'_p \in O_{X,p} \quad \text{such that} \quad g'_p w_o - h_p w_1 \in A_p .$$

Take representatives $g, g', h \in O_X(U)$ of g_p, g'_p, h_p in an open neighbourhood U of p such that

$$gw_o - hw_1,\ g'w_o - hw_1 \in A(U) .$$

Then we obtain

$$(g_x - g'_x)w_o \in A_x .$$

If we put $w_o = 1$ and if we assume $x \in U \setminus A$, then

$$g_x - g'_x = 0$$

since Γ is a holomorphic graph in x. Since A is analytically rare we conclude

$$g - g' = 0 \quad \text{and} \quad g_p = g'_p .$$

Now we define

$$f_p := \frac{g_p}{h_p} .$$

We have to show that these germs match together to a section $f \in M_X(X)$.
Assume that for $q \in X$ we have obtained

$$f_q = \frac{a_q}{b_q} .$$

We take open neighbourhoods U and V of p and q with representatives $g, h \in O_X(U)$ and $a, b \in O_X(V)$. Then we have

$$ah - bg = 0 \quad \text{on} \quad (U \cap V) \setminus A$$

since there Γ is a holomorphic graph. A being analytically rare the same is true on $U \cap V$.

Now it is very easy to check that the map

$$\Gamma \longmapsto f$$

obtained above is the inverse of our given map.
If f is obtained from Γ we construct the graph Γ_f.
Obviously Γ and Γ_f coincide outside $\sigma^{-1}(A)$. But then the dentity Theorem of 0.44 implies $\Gamma = \Gamma_f$.
Conversely take Γ_f, the graph of a meromorphic function f. Denote by f'

the meromorphic function extracted from Γ_f. Outside A we have

$$f = f' \;;$$

hence this is true everywhere (Remark 2 in 4.4).

Examples how this result can be applied will be given in 4.8 and 4.9.

4.7. A. HURWITZ [74] gave the first proof of the theorem that every meromorphic function on $P_n(\mathbb{C})$ is rational. This result had already been stated by WEIERSTRASS [149]. (See also [76]). W.L. CHOW [25] extended this theorem to projective algebraic varieties. In fact, this is an easy consequence of 4.3. It should be noted that a somewhat different short proof can be given with the methods of FAC [128].

Recall that, by definition, a *projective algebraic variety* is a closed complex subspace

$$X \hookrightarrow \mathbb{P}_n(\mathbb{C}) \;.$$

In order to define rational functions on such a space we use the following notations (see 1.3).

For every $\nu \in \{0,\ldots,n\}$ we consider the open subset

$$V_\nu := \{(z_o:\ldots:z_n) \in \mathbb{P}_n(\mathbb{C}) : z_\nu \neq 0\}$$

For every homogeneous polynomial

$$G \in \mathbb{C}[z_o,\ldots,z_n]$$

and a fixed ν we obtain an inhomogeneous polynomial

$$G_\nu \in \mathbb{C}[x_1,\ldots,x_n] \;, \quad \text{where} \quad x_1 = \frac{z_o}{z_\nu}, \ldots, x_\nu = \frac{z_{\nu-1}}{z_\nu}, \; x_{\nu+1} = \frac{z_{\nu+1}}{z_\nu}, \ldots, x_n = \frac{z_n}{z_\nu} \;.$$

G_ν may be considered as a polynomial function on V_ν as in 1.3.

If $X \hookrightarrow \mathbb{P}_n(\mathbb{C})$ is a projective algebraic variety, we consider the corresponding open subsets

$$X_\nu := X \cap V_\nu \subset X \;.$$

For G as above we denote by G'_ν the holomorphic function induced on X_ν by G_ν. Now a meromorphic function f on X is called *rational*, if there are homogeneous polynomials

$$G, H \in \mathbb{C}[z_o,\ldots,z_n]$$

of the same degree such that for every $\nu \in \{0,\ldots,n\}$ we have

$$f = \frac{G'_\nu}{H'_\nu} \quad \text{on } X_\nu$$

(this includes the condition that nowhere on X_ν the germ of H'_ν is a zerodivisor).

Theorem of HURWITZ. Let X be a reduced and irreducible projective algebraic variety. Then every meromorphic function on X is rational.

It should be noted that the restriction to the irreducible case is not essential but it helps to avoid some algebraic complications.

Proof. Assume we have $X \hookrightarrow \mathbb{P}_n(\mathbb{C})$. If f is meromorphic on X, we consider its graph

$$\Gamma \hookrightarrow X \times \mathbb{P}_1(\mathbb{C}) \hookrightarrow \mathbb{P}_n(\mathbb{C}) \times \mathbb{P}_1(\mathbb{C}) .$$

By the Theorem of CHOW (Corollary 2 in 4.3) there are polynomials

$$F_1,..,F_k \in \mathbb{C}[z_o,..,z_n;w_o,w_1] ,$$

homogeneous with respect to z_i and w_j, generating the ideal of Γ in $\mathbb{P}_n(\mathbb{C}) \times \mathbb{P}_1(\mathbb{C})$. We may assume that we have $\ell \le k$ such that

$$F_{\ell+1},...,F_k \in \mathbb{C}[z_o,..,z_n] ,$$

i.e. they have degree 0 with respect to w_j and they generate the ideal of X in $\mathbb{P}_n(\mathbb{C})$.

Now we claim that there is at least one polynomial

$$F \in \{F_1,..,F_\ell\}$$

such that F has degree 1 with respect to w_o,w_1 , i.e.

$$F = aw_o - bw_1 , \quad \text{where} \quad a,b \in \mathbb{C}[z_o,..,z_n]$$

are homogeneous. For this purpose we assume X is not contained in the hypersurface $x_o = 0$. Then the affine part $X_o \subset X$ with $z_o \neq 0$ is dense in X. In particular, there is a point $p \in X_o$ such that f is holomorphic in p.

Consider the restriction f_o of f to X_o, its graph

$$\Gamma_o \hookrightarrow X_o \times \mathbb{P}_1(\mathbb{C})$$

and the corresponding sheaf of homogeneous ideals

$$A \subset \mathcal{O}_{X_o}[w_o,w_1] .$$

It is generated by sections

$$\varphi_1,..,\varphi_\ell \in \mathcal{O}_{X_o}(X_o)[w_o,w_1]$$

which are induced by the polynomials

$$F_{1,o},..., F_{\ell,o} \in \mathbb{C}[x_1,..,x_n;w_o,w_1]$$

obtained from $F_1,..,F_\ell$ by putting $x_i = z_i/z_o$.

Since f is holomorphic in p, we have $\alpha,\beta \in \mathcal{O}_{X,p}$ with

$$\alpha w_o + \beta w_1 \in A_p \quad \text{and} \quad \beta(p) \neq 0$$

(see 4.5). This implies that there is at least one

$$\varphi = a_o w_o - b_o w_1 \in \{\varphi_1,\ldots,\varphi_\ell\}$$

with $b_o \neq 0$, and the corresponding F is as desired. Now it is easy to check that

$$f_o = \frac{a_o}{b_o}$$

and since X is irreducible this implies the analogous relation on the other affine parts.

4.8. In 2.23 we had proved the Riemann removable singularity theorems for holomorphic functions. For meromorphic functions this is much more difficult. It is a classical error that this problem may immediately be reduced to the holomorphic case by extending numerators and denominators (see e.g. [87]). In contrast to the extension theorems for holomorphic functions (see 2.27) it is not necessary to assume that the space is normal.

LEVIs removable singularity theorem. Let X be a reduced and irreducible complex space with an analytic subset $A \subset X$ of codimension at least 2. Then every meromorphic function

$$f \in M_X(X \setminus A)$$

has a unique extension $F \in M_X(X)$.

Proof (see [99], [107]). Consider the graph

$$\Gamma_f \hookrightarrow (X \setminus A) \times \mathbb{P}_1(\mathbb{C}) .$$

By using the Lemma of RITT it is easy to see that Γ_f is reduced (since this is obviously true outside the set of poles). The analytic set

$$B := \rho^{-1}(A) \subset X \times \mathbb{P}_1(\mathbb{C})$$

(ρ denotes the canonical projection) is of codimension 2, while Γ_f is of codimension 1. Hence the closure

$$\Gamma := \overline{\Gamma_f} \subset X \times \mathbb{P}_1(\mathbb{C})$$

is analytic by the REMMERT-STEIN-THULLEN Theorem ([146], [118], [107]). Since A is rare, Γ is a meromorphic graph which determines a meromorphic function F on X. This is the required extension.

If X is not reduced the REMMERT-STEIN-THULLEN Theorem may be replaced by extension theorems for sheaves of ideals ([147], [37]). This case is considered in [92].

<u>4.9.</u> In 4.2 we had defined proper modifications. By using 4.6 it is easy to see that they do not affect meromorphic functions.

<u>Proposition.</u> If $\varphi: X \to Y$ is a proper modification then the canonical homomorphism

$$\tilde{\varphi}: \mathcal{O}_Y \longrightarrow \varphi_*\mathcal{O}_X$$

is injective and extends to an isomorphism

$$\hat{\varphi}: \mathcal{M}_Y \longrightarrow \varphi_*\mathcal{M}_X .$$

Of course $\tilde{\varphi}$ need not be an isomorphism (take for example a normalization!

<u>Proof.</u> By assumption we have analytically rare sets $A \subset Y$ and $B := \varphi^{-1}(A) \subset X$ such that φ is biholomorphic outside A and B. Consider $p \in X$ and $q := \varphi(p) \in Y$. Since $A \subset Y$ is rare the homomorphism

$$\tilde{\varphi}_q: \mathcal{O}_{Y,q} \longrightarrow (\varphi_*\mathcal{O}_X)_q$$

is injective and since $B \subset X$ is also rare

$$\tilde{\varphi}_p: \mathcal{O}_{Y,q} \longrightarrow \mathcal{O}_{X,p}$$

is injective. For the multiplicatively closed system of non-zerodivisors we claim

$$\tilde{\varphi}_p(S_{Y,q}) \subset S_{X,p} . \qquad (*)$$

Assume $\alpha \in S_{X,q}$ is a non-zerodivisor. If $a_p := \tilde{\varphi}_p(\alpha)$ and $a_p \cdot b_p = 0$ for some $b_p \in \mathcal{O}_{X,p}$ then take an open neighbourhood $U \subset X$ of p and representatives $a, b \in \mathcal{O}_X(U)$ of a_p and b_p. If U is sufficiently small, then a_x is a non-zerodivisor for every $x \in U \setminus B$ since $\tilde{\varphi}_x$ is an isomorphism. Hence $b_x = 0$ for all $x \in U \setminus B$ and consequently $b = 0$. Therefore a_p is a non-zerodivisor.

Now $(*)$ implies that $\tilde{\varphi}_p$ inducces homomorphisms

$$\hat{\varphi}_p: \mathcal{M}_{Y,q} \longrightarrow \mathcal{M}_{X,p} \quad \text{and}$$

$$\hat{\varphi}: \mathcal{M}_Y \longrightarrow \varphi_*\mathcal{M}_X .$$

In order to show that $\hat{\varphi}$ is an isomorphism it is sufficient to construct an inverse of

$$\hat{\varphi}(Y): \mathcal{M}_Y(Y) \longrightarrow \mathcal{M}_X(X)$$

since for every open $V \subset Y$ and $U = \varphi^{-1}(V) \subset X$ the map

$$\varphi|U: U \longrightarrow V$$

is again a proper modification. So take $f \in M_X(X)$ and consider its graph

$$\Gamma_f \hookrightarrow X \times \mathbb{P}_1(\mathbb{C})$$

and the proper holomorphic map

$$\psi := \varphi \times id_{\mathbb{P}_1(\mathbb{C})}: X \times \mathbb{P}_1(\mathbb{C}) \longrightarrow Y \times \mathbb{P}_1(\mathbb{C}) .$$

We define

$$\Gamma := \psi(\Gamma_f) \hookrightarrow Y \times \mathbb{P}_1(\mathbb{C}) .$$

Γ is a holomorphic graph outside the analytically rare set $A \cup \varphi(P(f)) \subset Y$, hence a meromorphic graph. According to 4.5, Γ determines a meromorphic function f' on Y. This yields a map

$$M_X(X) \longrightarrow M_Y(Y), \quad f \longmapsto f' ,$$

and it is easy to check that it is the inverse of $\hat{\varphi}(Y)$.

Corollary. Let $\varphi: X \to Y$ be a proper modification which is finite. If Y is normal, then φ is biholomorphic.

An analogous (but much deeper) result in algebraic geometry is called ZARISKIs Main Theorem (see [153], [104], [53]).

Proof (see [94]). We have a commutative diagram

$$\begin{array}{ccc} M_Y & \xrightarrow{\hat{\varphi}} & \varphi_* M_X \\ \cup & & \cup \\ \mathcal{O}_Y & \xrightarrow{\tilde{\varphi}} & \varphi_* \mathcal{O}_X \end{array} ,$$

where $\hat{\varphi}$ is an isomorphism and $\tilde{\varphi}$ is a monomorphism. The Integrity Lemma of 2.27 implies that $\varphi_*\mathcal{O}_X$ is integral over $\mathcal{O}_Y$. Since Y is normal this implies that $\tilde{\varphi}$ is an isomorphism. By using Lemma 2 of 1.10 we conclude that φ is biholomorphic.

It is well known that there are complex manifolds of arbitrary dimension which have only constant meromorphic functions ([16], [91]). If X is such a manifold with dimension at least one and Y is a simple point then the holomorphic map

$$\varphi: X \to Y$$

is not a proper modification but the canonical map

$$M_Y(Y) \longrightarrow M_X(X)$$

is an isomorphism.

H. ISS'SA [75] proved the following

Theorem. Let X,Y be reduced and irreducible complex spaces and assume a monomorphism

$$\alpha: M_Y(Y) \longrightarrow M_X(X)$$

is given. If X is normal then

$$\alpha(\mathcal{O}_Y(Y)) \subset \mathcal{O}_X(X) .$$

If, in addition, Y is Stein then there is a unique holomorphic map

$$\varphi: X \to Y \quad \text{such that} \quad \alpha = \hat{\varphi} .$$

4.10. We conclude our notes with a proof of a theorem which has fascinated complex analysts for more than hundred years. For very precise historical and bibliographical notes we refer to [145].

Theorem of WEIERSTRASS-SIEGEL-THIMM. Let X be a compact complex space and assume we have meromorphic functions

$$f_1,\dots,f_n \in M_X(X) .$$

If $n > \dim X$, then they are *algebraically dependent*, i.e. there is a non zero polynomial $F \in \mathbb{C}[x_1,\dots,x_n]$ with

$$F(f_1,\dots,f_n) = 0 .$$

It immediately implies the

Corollary. If X is a reduced and irreducible compact complex space, then the transcendence degree of the field of meromorphic functions (see 4.4) can be estimated by

$$\operatorname{trdeg}_{\mathbb{C}} M_X(X) \leq \dim X .$$

For complex tori this result had been stated by WEIERSTRASS in 1869. After many attempts to handle the difficulties caused by the indeterminacies of the "functions" the first complete proofs were given in 1939 by SIEGEL [133] and THIMM [142]. A very elementary proof for complex manifolds was elaborated by SIEGEL [134]. He uses the well-known Lemma of H.A. SCHWARZ (see also BOCHNER [10] and SERRE [127]; SIEGELs proof can also be found in [145] and [132]). Furthermore there is a uniform estimate for the degrees of the polynomials F. In the corollary this implies that $M_X(X)$ is a finite algebraic extension of a rational function field. This becomes very clear by a method of STEIN [139], [140].

Here we reproduce a proof of REMMERT [115] which makes use of CHOWs Theorem.

<u>Proof of the Theorem of WEIERSTRASS-SIEGEL-THIMM.</u>

Given our meromorphic functions $f_1,..,f_n$ we consider their graphs and we define

$$\Gamma^* := \Gamma_{f_1} \times_X \cdots \times_X \Gamma_{f_n} \hookrightarrow X \times \mathbb{P}_1(\mathbb{C}) \times \cdots \times \mathbb{P}_1(\mathbb{C}) .$$

For $i=1,..,n$ the pole set $P(f_i) \subset X$ is analytically rare, hence

$$P := P(f_1) \cup \ldots \cup P(f_n) \subset X$$

is analytically rare. By the Lemma of RITT there is a unique closed complex subspace $\Gamma \hookrightarrow \Gamma^*$ such that the restriction

$$\sigma : \Gamma \to X$$

of the projection to the first factor is a proper modification, which is biholomorphic outside P. As an easy consequence of the Lemma of RITT we obtain

$$\dim \Gamma = \dim X .$$

Now we consider

$$\begin{array}{ccc} & \Gamma & \\ {\scriptstyle\sigma}\swarrow & & \searrow{\scriptstyle\pi} \\ X & & \mathbb{P}_1(\mathbb{C})\times\ldots\times\mathbb{P}_1(\mathbb{C}) \end{array} \qquad (*)$$

where π is the restriction of the other projection. Since X is compact, Γ is compact; in particular π is proper. Hence

$$\pi(\Gamma) \hookrightarrow \mathbb{P}_1(\mathbb{C})\times\ldots\times\mathbb{P}_1(\mathbb{C})$$

is a closed complex subspace with (see 1.18 and 3.8)

$$\dim \pi(\Gamma) \leq \dim X < n .$$

By CHOWs Theorem there is a non-zero polynomial

$$G \in \mathbb{C}[w_{1o}, w_{11}, w_{2o}, w_{21}, \ldots, w_{no}, w_{n1}]$$

which for every $i=1,..,n$ is homogeneous in w_{io}, w_{i1} (of some degree d_i) and contained in the ideal of $\pi(\Gamma)$. Associated to G there is an inhomogeneous polynomial $F \in \mathbb{C}[x_1,..,x_n]$ with

$$w_{1o}^{d_1} \cdot \ldots \cdot w_{no}^{d_n} \cdot F = G , \quad \text{where} \quad x_1 = \frac{w_{11}}{w_{1o}}, \ldots, x_n = \frac{w_{n1}}{w_{no}} .$$

If we restrict the diagram (*) to the complement of P we obtain

$$\begin{array}{ccc} & \Gamma' & \\ \sigma' \swarrow & & \searrow \pi' \\ X \setminus P & \xrightarrow{\varphi} & \mathbb{C}^n \end{array}$$

where $\varphi = \pi' \circ \sigma'^{-1}$. For i=1,..,n the functions

$$f_i' := f_i | (X \setminus P)$$

are holomorphic and by definition of φ we have (see 0.19)

$$\varphi = (f_1', .., f_n') .$$

By definition F is contained in the ideal of

$$\varphi(X \setminus P) = \pi'(\Gamma') \hookrightarrow \mathbb{C}^n ,$$

hence

$$F(f_1', .., f_n') = 0 .$$

This implies that the meromorphic function $F(f_1, .., f_n)$ vanishes outside P. Since $P \subset X$ is analytically rare we conclude (see 4.4)

$$F(f_1, .., f_n) = 0 .$$

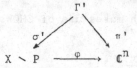

BIBLIOGRAPHY

[1] ABHYANKAR, S.S.: *Local Analytic Geometry*. Academic Press, New York-London 1964.

[2] AMAN, C.: *Globale Eigenschaften komplexer Räume*. Diplomarbeit, München 1975.

[3] ANDREIAN-CAZACU, C.: *Theorie der Funktionen mehrerer komplexer Veränderlicher*. Deutscher Verlag der Wissenschaften, Berlin 1975.

[4] ANDREOTTI, A.; STOLL, W.: *Analytic and Algebraic Dependence of Meromorphic Functions*. Lecture Notes in Mathematics Nr. 234. Springer Verlag, Berlin-Heidelberg-New York 1971.

[5] ATIYAH, M.; MACDONALD, I.G.: *Introduction to Commutative Algebra*. Addison-Wesley, Reading, Mass., 1969.

[6] BĂNICĂ, C.: *Un théomème concernant les familles analytiques d'espaces complexes*. Revue Roumaine 18, 2 (1973) 1515-1520.

[7] BĂNICĂ, C.; STĂNĂŞILĂ, O.: *Metode algebrice în teoria globală a spaţiilor complexe*. Editura Academici Republicii Socialiste România, Bucureşti 1974.

[8] BEHNKE, H.; THULLEN, P.: *Theorie der Funktionen mehrerer komplexer Veränderlichen*. Ergebnisse der Mathematik und ihrer Grenzgebiete, Band 51. Zweite, erweiterte Auflage. Springer-Verlag, Berlin-Heidelberg-New York 1970.

[9] BERGER, R.; KIEHL, R.; KUNZ, E.; NASTOLD, H.-J.: *Differentialrechnung in der analytischen Geometrie*. Lecture Notes in Mathematics, Nr. 38. Springer Verlag, Berlin-Heidolberg-New York 1967.

[10] BOCHNER, S.: *Algebraic and linear dependence of automorphic functions in several variables*. J. Indian Math. Soc. (N.S.) 16 (1952) 1-6.

[11] ——; MARTIN, W.T.: *Several Complex Variables*. Princeton, N.J. 1948.

[12] BOURBAKI, N.: *Topologie générale. Chap. I-II*. 2ième édition. Hermann, Paris 1951.

[13] ——: *Algèbre commutative. Chap. I-II*. Hermann, Paris 1961.

[14] ——: *Algèbre commutative. Chap. III-IV*. Hermann, Paris 1967.

[15] BRIESKORN, E.: *Singularitäten von Hyperflächen*. Vervielfältigtes Manuskript. Bonn, 1967.

[16] CALABI, E.; ECKMANN, B.: *A class of compact, complex manifolds which are not algebraic*. Ann. of Math. (2) 58 (1953) 494-500.

[17] CALABI, E.; ROSENLICHT, M.: *Complex analytic manifolds without countable base*. Proc. Amer. Math. Soc. 4 (1953) 335-340.

[18] CARTAN, H.: *Détermination des points exceptionnels d'un système de p fonctions analytiques de n variables complexes*. Bull. Sci. Math. 57 (1933) 334-344.

[19] ——: *Idéaux des fonctions analytiques de n variables complexes*. Ann. Sci. Ecole Norm. Sup. (3) 61 (1944) 149-197.

[20] ——: *Variétés analytiques et cohomologie*. Colloque sur les fonctions de plusieurs variables, tenu à Bruxelles, 1953. pp.41-55.

[21] ——: *Quotients of complex analytic spaces*. Contributions to function theory (Int. Coll. Function Theory, Bombay- 1960), 1-15, Tata Institute of Fundamental Research, Bombay, 1960.

[22] CARTAN, H.; EILENBERG, S.: *Homological Algebra*. Princeton University Press, Princeton, N.J. 1956.

[23] CARTAN, H.; SERRE, J.-P.: *Un théorème de finitude concernant les variétés analytiques compacts*. C.R. Acad. Sci. Paris 237 (1953) 128-130.

[24] CHEVALLEY, C.: *Theory of Lie Groups*. Princeton University Press, Princeton, N.J., 1946.

[25] CHOW, W.L.: *On compact analytic varieties*. Amer. J. Math. 71 (1949) 893-914.

[26] CLEMENTS, G.R.: *Implicit functions defined by equations with vanishing Jacobian*. Trans. Amer. Math. Soc. 14 (1913) 325-342.

[27] DOUADY, A.: *Le problème des modules pour les sous-espaces analytiques compacts d'un espace analytique donné*. Ann. Inst. Fourier 16 (1966) 1-98.

[28] ——: *Espaces analytiques sous-algébriques (d'après B.G. Moišezon)*. Séminaire Bourbaki, 20e année, 1967/68, No. 344.

[29] ——: *Flatness and privilege*. Enseignement math. 14 (1968) 47-74.

[30] FISCHER, G.: *Auflösung der Singularitäten gewisser holomorpher Abbildungen*. Bayer. Akad. Wiss. Math.-Naturw. Kl. S.-B. 1965, Abt. II, 75-85 (1966).

[31] ——: *Lineare Faserräume und kohärente Modulgarben über komplexen Räumen*. Arch. Math. 18 (1967) 609-617.

[32] ——: *Ein relativer Satz von Chow und die Elimination von Unbestimmtheitsstellen meromorpher Funktionen*. Math. Ann. 217 (1975) 145-152.

[33] FORSTER, O.: *Zur Theorie der Steinschen Algebren und Moduln*. Math. Z. 97 (1967) 376-405.

[34] ——; KNORR, K.: *Ein Beweis des Grauertschen Bildgarbensatzes nach Ideen von B. Malgrange*. Manuscripta Math. 5 (1971) 19-44.

[35] FRISCH, J.: *Points de platitude d'un morphisme d'espaces analytiques complexes*. Invent. math. 4 (1967) 118-138.

[36] ——: *Aplatissement en Géométrie analytique*. Ann. Sci. Ecole Norm. Sup. (4) 1 (1968) 305-312.

[37] ——; GUENOT, J.: *Prolongement des faisceaux analytiques cohérents*. Invent. math. 7 (1969) 321-343.

[38] GIESECKE, B.: *Simpliziale Zerlegung abzählbarer analytischer Räume*. Math. Z. 83 (1964) 177-213.

[39] GILMARTIN, M.: *Nondifferentiability of retractions of* $\mathbb{C}^n$ *to subvarieties*. Proc. Amer. Math. Soc. 16 (1965) 1028-1029.

[40] GODEMENT, R.: *Topologie algébrique et théorie des faisceaux*. Hermann, Paris, 1958.

[41] GRAUERT, H.: *Charakterisierung der holomorph vollständigen komplexen Räume*. Math. Ann. 129 (1955) 233-259.

[42] ——: *Ein Theorem der analytischen Garbentheorie und die Modulräume komplexer Strukturen*. Inst. Hautes Etudes Sci. Publ. Math. No. 5 (1960).

[43] ——: *Über Modifikationen und exzeptionelle analytische Mengen*. Math. Ann. 146 (1962) 331-368.

[44] GRAUERT, H.; FRITZSCHE, K.: *Einführung in die Funktionentheorie mehrerer Veränderlicher*. Springer-Verlag, Berlin-Heidelberg-New York 1974.

[45] ——; KERNER, H.: *Deformationen von Singularitäten komplexer Räume*. Math. Ann. 153 (1964) 236-260.

[46] ——; REMMERT, R.: *Zur Theorie der Modifikationen I. Stetige und eigentliche Modifikationen komplexer Räume*. Math. Ann. 129 (1955) 274-296.

[47] ——; ——: *Espaces analytiquement complets*. C.R. Acad. Sci. 245 (1957) 882-885.

[48] ——; ——: *Komplexe Räume*. Math. Ann. 136 (1958) 245-318.

[49] ——; ——: *Bilder und Urbilder analytischer Garben*. Ann. of Math. (2) 68 (1958) 393-443.

[50] ——; ——: *Analytische Stellenalgebren*. Die Grundlehren der mathematischen Wissenschaften, Band 176. Springer Verlag, Berlin-Heidelberg-New York 1971.

[51] ——; ——: *Steintheorie*. To appear.

[52] GROTHENDIECK, A.: *Techniques de construction en géométrie analytique*. Séminaire Henri Cartan, 13ième année (1960/61).

[53] ——: *Eléments de géométrie algébrique III. Etude cohomologique des faisceaux cohérents. I.* Rédigé avec la collaboration de J. Dieudonné. Inst. Hautes Etudes Sci. Publ. Math. No. 11 (1961).

[54] ——; DIEUDONNE, J.: *Eléments de géométrie algébrique*. Die Grundlehren der mathematischen Wissenschaften, Band 166. Springer-Verlag, Berlin-Heidelberg-New York 1971.

[55] GUNNING, R.C.: *Lectures on Complex Analytic Varieties. The Local Parametrization Theorem*. Princeton University Press, Princeton, N.J. 1970.

[56] ——: *Lectures on Complex Analytic Varieties. Finite Analytic Mappings*. Princeton University Press, Princeton, N.J. 1974.

[57] ——; ROSSI, H.: *Analytic Functions of Several Complex Variables*. Prentice Hall Inc., Englewood Cliffs, N.J. 1965.

[58] HAKIM, M.: *Topos annelés et schémas relatifs*. Ergebnisse der Mathematik und ihrer Grenzgebiete 64. Springer Verlag, Berlin-Heidelberg-New York 1972.

[59] HERMES, H.: *Analytische Mannigfaltigkeiten in Riemannschen Bereichen*. Math. Ann. 120 (1949) 539-562.

[60] HILBERT, D.: *Über die Theorie der algebraischen Formen*. Math. Ann. 36 (1890) 473-534.

[61] HIRONAKA, H.: *Resolution of singularities of an algebraic variety over a field of characteristic zero I*. Ann. of Math. (2) 79 (1964) 109-203.

[62] ——: *Resolution of singularities of an algebraic variety over a field of characteristic zero II*. Ann. of Math. (2) 79 (1964) 205-326.

[63] ——: *A fundamental lemma on point modifications*. Proc. Conference on Complex Analysis (Minneapolis, 1964), 194-215, Springer, Berlin-Heidelberg-New York 1965.

[64] ——: *Gardening of infinitely near singularities*. Proc. 5th Nordic Summer School in Mathematics, Oslo, August 5-25, 1970. Volters-

Noordhoff Publ. Gr. Netherland.

[65] HIRONAKA, H.: *Bimeromorphic smoothing of complex analytic spaces.* Preprint, University of Warwick 1971.

[66] ——: *Introduction to the theory of infinitely near singular points.* Memorias de matematica del instituto "Jorge Juan" 28. Madrid 1974.

[67] ——: *Flattening Theorem in Complex-Analytic Geometry.* Amer. J. Math. 97 (1975) 503-547.

[68] ——; ROSSI, H.: *On the equivalence of imbeddings of exceptional complex spaces.* Math. Ann. 156 (1964) 313-333.

[69] HIRZEBRUCH, F.: *Über vierdimensionale Riemannsche Flächen mehrdeutiger analytischer Funktionen von zwei komplexen Veränderlichen.* Math. Ann. 126 (1953) 1-22.

[70] HOLMANN, H.: *Local properties of holomorphic mappings.* Proc. Conf. Complex Analysis (Minneapolis, 1964) 94-109, Springer, Berlin 1965.

[71] HOPF, H.: *Schlichte Abbildungen und lokale Modifikationen 4-dimensionaler komplexer Mannigfaltigkeiten.* Comment. Math. Helv. 29 (1955) 132-156.

[72] HÖRMANDER, L.: *An introduction to complex analysis in several variables.* Princeton, N.J.; Van Nostrand 1966.

[73] HOUZEL, C.: *Géométrie analytique locale.* Séminaire Henri Cartan, 13ième année (1960/61).

[74] HURWITZ, A.: *Beweis des Satzes, daß eine einwertige Funktion beliebig vieler Variablen, welche überall als Quotient zweier Potenzreihen dargestellt werden kann, eine rationale Funktion ihrer Argumente ist.* J. Reine Angew. Math. 95 (1883) 201-206.

[75] ISS'SA, H.: *On the meromorphic function field of a Stein variety.* Ann. of Math. (2) 83 (1966) 34-46.

[76] JACKSON, D.: *Note on rational functions of several complex variables.* J. Reine Angew. Math. 146 (1916) 185-188.

[77] JURCHESCU, M.: *On a theorem of Stoilow.* Math. Ann. 138 (1959) 332-334.

[78] KAUP, B.: *Äquivalenzrelationen auf allgemeinen komplexen Räumen.* Schr. Math. Inst. Univ. Münster Nr. 39 (1968).

[79] ——: *Über Kokerne und Pushouts in der Kategorie der komplex-analytischen Räume.* Math. Ann. 189 (1970) 60-76.

[80] ——: *Ein Kriterium für platte holomorphe Abbildungen.* Bayer. Akad. Wiss. Math.-Naturw. Kl. S.-B. 1968, Abt. II, 101-105 (1969).

[81] KERNER, H.: *Bemerkung zu einem Satz von H. Grauert und R. Remmert.* Math. Ann. 157 (1964), 206-209.

[82] ——: *Zur Theorie der Deformationen komplexer Räume.* Math. Z. 103 (1968) 389-398.

[83] KIEHL, R.: *Note zu der Arbeit von J. Frisch: "Points de platitude d'un morphisme d'espaces analytiques complexes.* Invent. Math. 4 (1967) 139-141.

[84] ——: *Analytische Familien affinoider Algebren.* S.-B. Heidelberger Akad. Wiss. Math.-Naturw. Kl. 1968, 23-49.

[85] ——: *Äquivalenzrelationen in analytischen Räumen.* Math. Z. 105 (1968) 1-20.

[86] KIEHL, R.; VERDIER, J.L.: *Ein einfacher Beweis des Kohärenzsatzes von Grauert.* Math. Ann. 195 (1971/72) 24-50.

[87] KISTLER, H.: *Über Funktionen von mehreren komplexen Veränderlichen.* Dissertation, Göttingen 1905.

[88] KNESER, H.: *Analytische Mannigfaltigkeiten im komplexen projektiven Raum.* Math. Nachr. 4 (1950/51) 382-391.

[89] KNESER, M.: *Beispiel einer dimensionserhöhenden analytischen Abbildung zwischen überabzählbaren Mannigfaltigkeiten.* Arch. Math. 11 (1960) 280-281.

[90] KNORR, K.; SCHNEIDER, M.: *Relativexzeptionelle analytische Mengen.* Math. Ann. 193 (1971) 238-254.

[91] KOPFERMANN, K.: *Maximale Untergruppen abelscher komplexer Liescher Gruppen.* Schr. Math. Inst. Univ. Münster Nr. 29 (1964).

[92] KRAUS, G.: *Holomorphe Korrespondenzen und meromorphe Abbildungen allgemeiner komplexer Räume - Fortsetzungssätze.* Manuscripta math. 6 (1972) 1-15.

[93] ——: *Meromorphe Funktionen auf allgemeinen komplexen Räumen.* Math. Ann. 209 (1974) 257-265.

[94] ——: *Beiträge zur Theorie der allgemeinen komplexen Räume.* Manuskript, München 1976.

[95] KUHLMANN, N.: *Über die Auflösung der Singularitäten 3-dimensionaler komplexer Räume. I.* Math. Ann. 151 (1963) 304-331.

[96] ——: *Über die Auflösung der Singularitäten 3-dimensionaler komplexer Räume. II.* Math. Ann. 154 (1964) 387-405.

[97] ——: *Über holomorphe Abbildungen komplexer Räume.* Arch. Math. 15 (1964) 81-90.

[98] ——: *Über die Reinheit von Entartungs- und Verzweigungsmengen.* Math. Ann. 178 (1968) 25-43.

[99] LEVI, E.E.: *Studii sui punti singolari essenziali delli funzioni analitiche di due o più variabili complesse.* Ann. Mat. Pura Appl. (3) 17 (1910) 61-87.

[100] ŁOJASIEWICZ, S.: *Triangulation of semi-analytic sets.* Ann. Scuola Norm. Sup. Pisa (3) 18 (1964) 449-474.

[101] MALGRANGE, B.: *Analytic Spaces.* Enseignement Math. (2) 14 (1968) 1-28.

[102] MARKOE, A.: *A characterization of normal analytic spaces by the homological codimension of the structure sheaf.* Pacific J. Math. 52 (1974) 485-489.

[103] MOIŠEZON, B.G.: *Modifications of complex varieties and the Chow Lemma.* Classification of algebraic varieties and compact complex manifolds. Edited by H. Popp, pp. 133-139. Lecture Notes in Mathematics, Vol. 412. Springer-Verlag, Berlin-Heidelberg-New York 1975.

[104] MUMFORD, D.: *Introduction to Algebraic Geometry. Preliminary version of the three first chapters.* Mimeographed Notes.

[105] NAGATA, M.: *Local Rings.* Interscience, New York 1962.

[106] NARASIMHAN, R.: *A note on Stein spaces and their normalization.* Ann. Scuola Norm. Sup. Pisa (3) 16 (1962) 327-333.

[107] NARASIMHAN, R.: *Introduction to the Theory of Analytic Spaces.* Lecture Notes in Mathematics, No. 25. Springer-Verlag, Berlin-Heidelberg-New York 1966.

[108] ——: *Several Complex Variables.* The University of Chicago Press, Chicago-London 1971.

[109] OORT, F.: *Commutative Group Schemes.* Lecture Notes in Mathematics Vol. 15. Springer-Verlag, Berlin-Heidelberg-New York, 1963.

[110] ——: *Algebraic group schemes in characteristic zero are reduced.* Invent. Math. 2 (1966/67) 79-80.

[111] OSGOOD, W.F.: *Lehrbuch der Funktionentheorie II.1.* 2. Aufl. B.G. Teubner, Leipzig-Berlin 1929.

[112] PAREIGIS, B.: *Kategorien und Funktoren.* Mathematische Leitfäden. B.G. Teubner, Stuttgart 1969.

[113] PRILL, D.: *Über lineare Faserräume und schwach negative Geradenbündel.* Math. Z. 105 (1968) 313-326.

[114] REIFFEN, H.-J.; VETTER, U.: *Pfaffsche Formen auf komplexen Räumen.* Math. Ann. 167 (1966) 338-350.

[115] REMMERT, R.: *Meromorphe Funktionen in kompakten komplexen Räumen.* Math. Ann. 132 (1956) 277-288.

[116] ——: *Sur les espaces analytiques holomorphiquement séparables et holomorphiquement convexes.* C.R. Acad. Sci. Paris 243 (1956) 118-121.

[117] ——: *Holomorphe und meromorphe Abbildungen komplexer Räume.* Math. Ann. 133 (1957) 328-370.

[118] ——; STEIN, K.: *Über die wesentlichen Singularitäten analytischer Mengen.* Math. Ann. 126 (1953) 263-306.

[119] ——; ——: *Eigentliche holomorphe Abbildungen.* Math. Z. 73 (1960) 159-189.

[120] RITT, J.F.: *Differential equations from the algebraic standpoint.* (AMS coll. publ., vol. 14) New York: AMS 1932.

[121] ROSSI, H.: *Vector fields on analytic spaces.* Ann. of Math. (2) 78 (1963) 455-467.

[122] SCHEJA, G.: *Fortsetzungssätze der komplex-analytischen Cohomologie und ihre algebraische Charakterisierung.* Math. Ann. 157 (1964) 75-94.

[123] ——: *Eine Anwendung Riemannscher Hebbarkeitssätze für analytische Cohomologieklassen.* Arch. Math. 12 (1961) 341-348.

[124] SCHUMACHER, G.: *Über ebene regulär faktorisierbare und einfache holomorphe Abbildungen.* Manuscripta math. 15 (1975) 33-43.

[125] ——: *Über regulär faktorisierbare holomorphe Abbildungen und ein topologisches Regularitätskriterium.* Math. Ann., to appear.

[126] SCHUSTER, H.W.: *Infinitesimale Erweiterungen komplexer Räume.* Comment. Math. Helv. 45 (1970) 265-286.

[127] SERRE, J.P.: *Fonctions automorphes: Quelques majorations dans le cas où X/G est compact.* Séminaire Cartan 1953/54, Exposé II.

[128] ——: *Faisceaux algébriques cohérents.* Ann. of Math. (2) 61 (1955) 197-278.

[129] SERRE, J.P.: *Géométrie algébrique et géométrie analytique.* Ann. Inst. Fourier 6 (1955) 1-42.

[130] ——: *Algèbre locale. Multiplicités.* Cours au Collège de France 1957-1958, rédigé par Pierre Gabriel. Seconde Edition, 1965. Lecture Notes in Mathematics, Vol. 11. Springer-Verlag, Berlin-Heidelberg-New York 1965.

[131] ——: *Prolongements des faisceaux analytiques cohérents.* Ann. Inst. Fourier (Grenoble) 16 (1966) fasc. 1, 363-374.

[132] SHAFAREVICH, I.R.: *Basic Algebraic Geometry.* Die Grundlehren der mathematischen Wissenschaften, Band 213. Springer-Verlag, Berlin-Heidelberg-New York 1974.

[133] SIEGEL, C.L.: *Einführung in die Theorie der Modulfunktionen n-ten Grades.* Math. Ann. 116 (1939) 617-657.

[134] ——: *Meromorphe Funktionen auf kompakten analytischen Mannigfaltigkeiten.* Nachr. Akad. Wiss. Göttingen. Math. Phys. Kl. IIa. 1955, 71-77.

[135] SIMHA, R.R.: *Über die kritischen Werte gewisser holomorpher Abbildungen.* Manuscripta math. 3 (1970) 97-104.

[136] SIU, Y.T.: *Noether-Lasker decomposition of coherent subsheaves.* Trans. Amer. Math. Soc. 135 (1969) 375-385.

[137] ——; TRAUTMANN, G.: *Gap-sheaves and Extension of Coherent Analytic Subsheaves.* Lecture Notes in Mathematics, Vol. 172. Springer Verlag, Berlin-Heidelberg-New York 1971.

[138] STEIN, K.: *Analytische Zerlegungen komplexer Räume.* Math. Ann. 132 (1956) 63-93.

[139] ——; *Maximale holomorphe und meromorphe Abbildungen I.* Amer. J. Math. 85 (1963) 298-313.

[140] ——: *Maximale holomorphe und meromorphe Abbildungen II.* Amer. J. Math. 86 (1964) 823-868.

[141] STORCH, U.: *Über starre analytische Algebren.* Math. Ann. 179 (1968) 53-62.

[142] THIMM, W.: *Über algebraische Relationen zwischen meromorphen Funktionen in abgeschlossenen Räumen.* Dissertation, Königsberg 1939.

[143] ——: *Untersuchungen über das Spurproblem von holomorphen Funktionen auf analytischen Mengen.* Math. Ann. 139 (1959) 95-114.

[144] ——: *Lückengarben von kohärenten analytischen Modulgarben.* Math. Ann. 148 (1962) 372-394.

[145] ——: *Der Weierstraßsche Satz der algebraischen Abhängigkeit von abelschen Funktionen und seine Verallgemeinerungen.* Festschrift Gedächtnisfeier K. Weierstraß, pp. 123-154. Westdeutscher Verlag, Köln 1966.

[146] THULLEN, P.: *Über die wesentlichen Singularitäten analytischer Funktionen und Flächen im Raume von n komplexen Veränderlichen.* Math. Ann. 111 (1935) 137-157.

[147] TRAUTMANN, G.: *Ein Kontinuitätssatz für die Fortsetzung kohärenter analytischer Garben.* Arch. Math. 18 (1967) 188-196.

[148] WAERDEN, B.L. van der: *Zur algebraischen Geometrie III.* Math. Ann. 108 (1933) 694-698.

[149] WEIERSTRASS, K.: *Untersuchungen über die 2r-fachen periodischen Funktionen von r Veränderlichen*. J. Reine Angew. Math. 89 (1880) 1-8.

[150] WHITHEY, H.: *Complex Analytic Varieties*. Addison-Wesley, Reading, Mass. 1972.

[151] WIEGMANN, K.-W.: *Über Quotienten holomorph-konvexer komplexer Räume*. Math. Z. 97 (1967) 251-258.

[152] ——: *Strukturen auf Quotienten komplexer Räume*. Comment. Math. Helv. 44 (1969) 93-116.

[153] ZARISKI, O.: *Foundations of a general theory of birational correspondences*. Trans. Amer. Math. Soc. 53 (1943) 490-542.

[154] ——: *Reduction of singularities of algebraic three-dimensional varieties*. Ann. of Math. 45 (1944) 472-542.

[155] ——: *On the purity of the branch locus of algebraic functions*. Proc. Nat. Acad. USA 44 (1957) 791-796.

[156] ——: *Studies in equisingularily I*. Amer. J. Math. 87 (1965) 507-536.

[157] ——; SAMUEL, P.: *Commutative Algebra*. Van Nostrand, Princeton, N.J. 1960.

INDEX

Vol. 399: Functional Analysis and its Applications. Proceedings 1973. Edited by H. G. Garnir, K. R. Unni and J. H. Williamson. II, 584 pages. 1974.

Vol. 400: A Crash Course on Kleinian Groups. Proceedings 1974. Edited by L. Bers and I. Kra. VII, 130 pages. 1974.

Vol. 401: M. F. Atiyah, Elliptic Operators and Compact Groups. V, 93 pages. 1974.

Vol. 402: M. Waldschmidt, Nombres Transcendants. VIII, 277 pages. 1974.

Vol. 403: Combinatorial Mathematics. Proceedings 1972. Edited by D. A. Holton. VIII, 148 pages. 1974.

Vol. 404: Théorie du Potentiel et Analyse Harmonique. Edité par J. Faraut. V, 245 pages. 1974.

Vol. 405: K. J. Devlin and H. Johnsbråten, The Souslin Problem. VIII, 132 pages. 1974.

Vol. 406: Graphs and Combinatorics. Proceedings 1973. Edited by R. A. Bari and F. Harary. VIII, 355 pages. 1974.

Vol. 407: P. Berthelot, Cohomologie Cristalline des Schémas de Caracteristique $p > o$. II, 604 pages. 1974.

Vol. 408: J. Wermer, Potential Theory. VIII, 146 pages. 1974.

Vol. 409: Fonctions de Plusieurs Variables Complexes, Seminaire François Norguet 1970–1973. XIII, 612 pages. 1974.

Vol. 410: Séminaire Pierre Lelong (Analyse) Année 1972–1973. VI, 181 pages. 1974.

Vol. 411: Hypergraph Seminar. Ohio State University, 1972. Edited by C. Berge and D. Ray-Chaudhuri. IX, 287 pages. 1974.

Vol. 412: Classification of Algebraic Varieties and Compact Complex Manifolds. Proceedings 1974. Edited by H. Popp. V, 333 pages. 1974.

Vol. 413: M. Bruneau, Variation Totale d'une Fonction. XIV, 332 pages. 1974.

Vol. 414: T. Kambayashi, M. Miyanishi and M. Takeuchi, Unipotent Algebraic Groups. VI, 165 pages. 1974.

Vol. 415: Ordinary and Partial Differential Equations. Proceedings 1974. XVII, 447 pages. 1974.

Vol. 416: M. E. Taylor, Pseudo Differential Operators. IV, 155 pages. 1974.

Vol. 417: H. H. Keller, Differential Calculus in Locally Convex Spaces. XVI, 131 pages. 1974.

Vol. 418: Localization in Group Theory and Homotopy Theory and Related Topics. Battelle Seattle 1974 Seminar. Edited by P. J. Hilton. VI, 172 pages 1974.

Vol. 419: Topics in Analysis. Proceedings 1970. Edited by O. E. Lehto, I. S. Louhivaara, and R. H. Nevanlinna. XIII, 392 pages. 1974.

Vol. 420: Category Seminar. Proceedings 1972/73. Edited by G. M. Kelly. VI, 375 pages. 1974.

Vol. 421: V. Poénaru, Groupes Discrets. VI, 216 pages. 1974.

Vol. 422: J.-M. Lemaire, Algèbres Connexes et Homologie des Espaces de Lacets. XIV, 133 pages. 1974.

Vol. 423: S. S. Abhyankar and A. M. Sathaye, Geometric Theory of Algebraic Space Curves. XIV, 302 pages. 1974.

Vol. 424: L. Weiss and J. Wolfowitz, Maximum Probability Estimators and Related Topics. V, 106 pages. 1974.

Vol. 425: P. R. Chernoff and J. E. Marsden, Properties of Infinite Dimensional Hamiltonian Systems. IV, 160 pages. 1974.

Vol. 426: M. L. Silverstein, Symmetric Markov Processes. X, 287 pages. 1974.

Vol. 427: H. Omori, Infinite Dimensional Lie Transformation Groups. XII, 149 pages. 1974.

Vol. 428: Algebraic and Geometrical Methods in Topology, Proceedings 1973. Edited by L. F. McAuley. XI, 280 pages. 1974.

Vol. 429: L. Cohn, Analytic Theory of the Harish-Chandra C-Function. III, 154 pages. 1974.

Vol. 430: Constructive and Computational Methods for Differential and Integral Equations. Proceedings 1974. Edited by D. L. Colton and R. P. Gilbert. VII, 476 pages. 1974.

Vol. 431: Séminaire Bourbaki – vol. 1973/74. Exposés 436–452. IV, 347 pages. 1975.

Vol. 432: R. P. Pflug, Holomorphiegebiete, pseudokonvexe Gebiete und das Levi-Problem. VI, 210 Seiten. 1975.

Vol. 433: W. G. Faris, Self-Adjoint Operators. VII, 115 pages. 1975.

Vol. 434: P. Brenner, V. Thomée, and L. B. Wahlbin, Besov Spaces and Applications to Difference Methods for Initial Value Problems. II, 154 pages. 1975.

Vol. 435: C. F. Dunkl and D. E. Ramirez, Representations of Commutative Semitopological Semigroups. VI, 181 pages. 1975.

Vol. 436: L. Auslander and R. Tolimieri, Abelian Harmonic Analysis, Theta Functions and Function Algebras on a Nilmanifold. V, 99 pages. 1975.

Vol. 437: D. W. Masser, Elliptic Functions and Transcendence. XIV, 143 pages. 1975.

Vol. 438: Geometric Topology. Proceedings 1974. Edited by L. C. Glaser and T. B. Rushing. X, 459 pages. 1975.

Vol. 439: K. Ueno, Classification Theory of Algebraic Varieties and Compact Complex Spaces. XIX, 278 pages. 1975

Vol. 440: R. K. Getoor, Markov Processes: Ray Processes and Right Processes. V, 118 pages. 1975.

Vol. 441: N. Jacobson, PI-Algebras. An Introduction. V, 115 pages. 1975.

Vol. 442: C. H. Wilcox, Scattering Theory for the d'Alembert Equation in Exterior Domains. III, 184 pages. 1975.

Vol. 443: M. Lazard, Commutative Formal Groups. II, 236 pages. 1975.

Vol. 444: F. van Oystaeyen, Prime Spectra in Non-Commutative Algebra. V, 128 pages. 1975.

Vol. 445: Model Theory and Topoi. Edited by F. W. Lawvere, C. Maurer, and G. C. Wraith. III, 354 pages. 1975.

Vol. 446: Partial Differential Equations and Related Topics. Proceedings 1974. Edited by J. A. Goldstein. IV, 389 pages. 1975.

Vol. 447: S. Toledo, Tableau Systems for First Order Number Theory and Certain Higher Order Theories. III, 339 pages. 1975.

Vol. 448: Spectral Theory and Differential Equations. Proceedings 1974. Edited by W. N. Everitt. XII, 321 pages. 1975.

Vol. 449: Hyperfunctions and Theoretical Physics. Proceedings 1973. Edited by F. Pham. IV, 218 pages. 1975.

Vol. 450: Algebra and Logic. Proceedings 1974. Edited by J. N. Crossley. VIII, 307 pages. 1975.

Vol. 451: Probabilistic Methods in Differential Equations. Proceedings 1974. Edited by M. A. Pinsky. VII, 162 pages. 1975.

Vol. 452: Combinatorial Mathematics III. Proceedings 1974. Edited by Anne Penfold Street and W. D. Wallis. IX, 233 pages. 1975.

Vol. 453: Logic Colloquium. Symposium on Logic Held at Boston, 1972–73. Edited by R. Parikh. IV, 251 pages. 1975.

Vol. 454: J. Hirschfeld and W. H. Wheeler, Forcing, Arithmetic, Division Rings. VII, 266 pages. 1975.

Vol. 455: H. Kraft, Kommutative algebraische Gruppen und Ringe. III, 163 Seiten. 1975.

Vol. 456: R. M. Fossum, P. A. Griffith, and I. Reiten, Trivial Extensions of Abelian Categories. Homological Algebra of Trivial Extensions of Abelian Categories with Applications to Ring Theory. XI, 122 pages. 1975.